THEORY
OF FUNCTIONS

By
DR. KONRAD KNOPP
Professor of Mathematics at the University of Tübingen

Translated by
FREDERICK BAGEMIHL, M. A.
Instructor in Mathematics at the University of Rochester

PART TWO
APPLICATIONS AND CONTINUATION OF THE
GENERAL THEORY

NEW YORK
DOVER PUBLICATIONS

International Standard Book Number: 0-486-60157-9
Library of Congress Catalog Card Number: 45-6381

Manufactured in the United States of America
Dover Publications, Inc.
180 Varick Street
New York, N.Y. 10014

CONTENTS

SECTION I

Single-valued Functions

CHAPTER 1. ENTIRE FUNCTIONS

CHAPTER 2. MEROMORPHIC FUNCTIONS

CHAPTER 3. PERIODIC FUNCTIONS

SECTION II

Multiple-valued Functions

INTRODUCTION

The foundations of the general theory of analytic functions were laid in Part I of this *Theory of Functions*.[1] Special functions (such as e^z, sin z, log z, $\sqrt[v]{z}$, and others) or classes of functions (such as the rational or the entire functions) were dealt with there only occasionally. Now such more detailed investigations will come in greater measure to the foreground. Only once again, later on, will more general considerations be carried out, in order to clarify the situation left undiscussed in I, §24, pp. 103-104. In doing so, it will become apparent that the distinction between single-valued and multiple-valued functions which was indicated there is quite fundamental. This distinction will therefore serve from the outset as a standard for all of the following presentation.

From these two main classes we shall select several especially characteristic and important types of functions. A certain arbitrariness is unavoidable in this connection, since completeness within the close compass of this little book is naturally denied us. We shall get away from this danger most easily if we start with the elementary functions (the entire and fractional rational functions, e^z, sin z, log z, $\sqrt[v]{z}$, $\cdots$) as the most

[1]*Theory of Functions, Part I: Elements of the General Theory of Analytic Functions*, translated from the 5th German edition, New York, 1945,—referred to in the following, briefly, as "I", together with paragraph or page number.

important ones, and try to understand that which is essential and of a universal character in their principal properties.[1]

The *entire rational functions* (*polynomials*)—evidently the simplest and most transparent functions—are characterized (cf. I, p. 137) "purely function-theoretically" by the fact that they are regular in the entire plane and have a pole at the point ∞. If one ignores the last property, one arrives at the more general class of *entire functions*, which are characterized solely by the property of being regular in the entire plane (excluding ∞), and to which the entire rational and the entire transcendental functions belong as special cases. They also appeared to us in I, §27 to be the simplest, because their power-series expansion for an arbitrary center converges, and therefore represents the function, in the entire plane. Since analytic continuation, then, is out of the question, the entire functions are naturally single-valued. In their totality they are identical with the totality of everywhere-convergent power series of the form

$$g(z) = \sum_{n=0}^{\infty} a_n z^n,$$

and, as such, appear to be an immediate generalization of the entire rational functions.

[1]Thus, we can be concerned in the following with a selection only—with samples, so to speak. The theory of functions is a large realm, which cannot be explored on a journey of one or even several days. If, in spite of this, we undertake on the following pages to sketch briefly a few of the principal places in this realm, we must emphatically caution the reader not to identify the extent of this little volume with that of the theory of functions.

In the first chapter we shall approach these functions with the question: Which of the fundamental properties of the entire rational functions does the class of entire functions still possess, and which not?—and shall give several answers to this question.

According to I, §35, Theorems 1 and 2, the *fractional rational functions* are completely characterized from the purely function-theoretical point of view by the fact that they have no singularities other than poles in the entire plane and at the point ∞. If, here too, one ignores the last property concerning the point ∞, one arrives again at a more general class of functions, the so-called *meromorphic functions*, which are characterized solely by the property of having no singularities other than poles in the entire plane (excluding ∞).

In the second chapter we shall approach these functions, which will also prove to be single-valued, with the question to be formulated analogous to the one above.

The property of the functions e^z, sin z, and others, which is the most interesting from the function-theoretical standpoint, is their periodicity. In the third chapter we shall detach this property from the special nature of these functions, and investigate it more closely and purely function-theoretically. We thus arrive at the classes of *simply periodic* and *doubly periodic functions*. In the latter class we then meet, in particular, the *elliptic functions*.

These types chosen from the realm of single-valued functions will have to suffice.

In connection with *multiple-valued functions*, we shall first be concerned with sifting out the concept of the same more clearly than was possible in I, §24, and with giving a clearer notion of the essence of multiple-valuedness. This is accomplished in the fourth chapter by means of a very simple and, just for this reason, remarkably ingenious idea, that of the so-called *Riemann surface*. The construction of these surfaces is illustrated using the simplest multiple-valued functions,

$$\sqrt[p]{z}, \ \log z, \ \sqrt{(z - a_1)(z - a_2) \cdots (z - a_k)}.$$

In the fifth chapter, a particularly important and therefore also especially well-investigated class of multiple-valued functions, the class of *algebraic functions*, is treated in somewhat greater detail.

With the aid of the concept of the *algebraic singularity* acquired hereby, all gaps still contained in our definition of the complete analytic function or of the analytic configuration, given in I, pp.102-103, are finally filled in chapter 6. Thus is obtained in its complete generality the notion of the *analytic configuration*. This admirable concept, occupying the center of our considerations from the very beginning, though by no means to be mastered in the first attack, is undisputedly one of the most profound and beautiful in all of mathematical science.

SINGLE-VALUED FUNCTIONS

CHAPTER 1

ENTIRE FUNCTIONS

§1. Weierstrass's Factor-Theorem

The most important property of the entire rational functions is expressed in the fundamental theorem of algebra (see I, pp.113 and 139): *Every non-constant entire rational function has zeros.* Since e^z, for example, has no zeros (because $e^z \cdot e^{-z} = 1$), the question formulated above seems immediately doomed to unfruitfulness. Upon further investigation of the core of the matter, however, we shall see that this is not so. Indeed, if

$$g_0(z) = a_0 + a_1 z + \cdots + a_m z^m,$$

$$(m \geq 1, \, a_m \neq 0)$$

is an arbitrary, non-constant entire rational function, then it follows from the fundamental theorem of algebra that $g_0(z)$ can be written in the form

$$(1) \qquad g_0(z) = a_m (z - z_1)^{\alpha_1} (z - z_2)^{\alpha_2} \cdots (z - z_k)^{\alpha_k},$$

where $z_1, z_2, \cdots, z_k$ denote all the distinct zeros of $g_0(z)$, and $\alpha_1, \alpha_2, \cdots, \alpha_k$ denote their respective orders. We express this as follows:

(A) For every entire rational function there is a

1

so-called *factor representation*, which displays its zeros as to position and order.[1]

We infer immediately from this representation, that every other entire rational function $g(z)$ which has the same zeros to the same respective orders can differ from $g_0(z)$ only in the factor a_m. Furthermore, one can give these zeros any position and any order. In other words:

(B) It is always possible to construct an entire rational function whose zeros (finite in number, of course) are prescribed as to position and order. This function can be represented as a *product* which displays these zeros. The most general function of this kind is obtained from a particular one by multiplying it by an arbitrary non-zero factor ("a multiplicative entire rational function with no zeros").

If we take these two statements (A) and (B) to express the content of the fundamental theorem of algebra, then we shall see that all this can be carried over word for word to arbitrary entire functions.

To this end, we begin by setting ourselves the following **problem,** which corresponds to (B) and is fundamental for all that follows. We propose to investigate whether, and how, one can construct entire functions with prescribed zeros,[2] and to what extent an entire function is determined by these conditions.

[1]This also holds for entire rational functions "with no zeros", i.e., of degree zero (namely, the non-zero constants), for which the factor representation consists of the factor $a_m(=a_0 \neq 0)$ alone. On the other hand, our considerations, of course, no longer apply to the constant 0.

[2]This means that the function is to have zeros of certain prescribed orders at certain prescribed points, *and be distinct from zero at all other points.*

Entire functions with no zeros. Suppose the entire function to be constructed is to have no zeros at all. Then the constant 1, or the function e^z, or e^{z^2}, or, more generally, $e^{h(z)}$ is a solution of the problem, if $h(z)$ is a completely arbitrary *entire* function. The last answer is also the *most general* solution of the problem. That is, $e^{h(z)}$ (with $h(z)$ an arbitrary entire function) is not only always an entire function with no zeros, but conversely, every such function can be written in the form $e^{h(z)}$. We state this more briefly:

Theorem 1. *If $h(z)$ denotes an arbitrary entire function, then $e^{h(z)}$ is the most general entire function with no zeros.*[1]

Proof: We have only to show that if $H(z) = a_0 + a_1 z + a_2 z^2 + a_3 z^3 + \cdots$ is a given entire function with no zeros, another entire function $h(z) = b_0 + b_1 z + \cdots$ can be determined such that $e^{h(z)} = H(z)$. Now, since $H(z) \neq 0$, we have in particular $a_0 = H(0) \neq 0$. Hence, b_0 can be chosen such that $e^{b_0} = a_0$; for, e^z takes on every value except zero. Likewise, $\dfrac{1}{H(z)}$ is everywhere single-valued and regular, and is therefore

---

[1] This theorem seems almost trivial if we make use of the multiple-valued function log. For, let $H(z)$ be an entire function which is nowhere equal to zero. Then $h(z) = \log H(z)$, with the condition that, e.g., $h(0)$ be the principal value of $\log H(0)$, is also a regular function of z in a certain neighborhood of the origin. Its expansion there, $h(z) = b_0 + b_1 z + b_2 z^2 + \cdots$, consequently has a positive radius of convergence. This must (by I, §24, Theorem 1) be $+\infty$; because $\log H(z)$ can be singular only where $H(z)$ is singular or equal to zero, and hence, nowhere in the finite part of the plane.

an entire function. The same is true of $H'(z)$, so that

$$\frac{H'(z)}{H(z)} = c_0 + c_1 z + c_2 z^2 + \cdots$$

is also an entire function, and this series is everywhere convergent. The latter also holds for the series

$$b_0 + c_0 z + \frac{c_1}{2} z^2 + \cdots + \frac{c_{n-1}}{n} z^n + \cdots$$

$$= b_0 + b_1 z + \cdots + b_n z^n + \cdots ,$$

which accordingly represents an entire function, $h(z)$. If we set $e^{h(z)} = H_1(z)$, then

$$\frac{H_1'(z)}{H_1(z)} = c_0 + c_1 z + c_2 z^2 + \cdots = \frac{H'(z)}{H(z)},$$

and hence $H_1 \cdot H' - H \cdot H_1' = 0$. Consequently

$$\frac{H \cdot H_1' - H_1 \cdot H'}{H^2} = \frac{d}{dz}\left(\frac{H_1(z)}{H(z)}\right) = 0,$$

and the quotient of the two functions $H_1(z)$ and $H(z)$ is constant. For $z = 0$ we find the value 1 for this constant. Therefore

$$H(z) = H_1(z) = e^{h(z)}, \qquad \text{Q. E. D.}[1]$$

Having thus completely solved our problem for the case that *no* zeros are prescribed, it is easy to see the

[1]The proof actually demonstrates the following: If two functions $f(z)$ and $f_1(z)$ are *single-valued regular, and distinct from zero* in a region $\mathfrak{G}$, and if their logarithmic derivatives f'/f and f_1'/f_1 coincide there, then they differ in $\mathfrak{G}$ by at most a constant factor (which must, of course, be equal to unity if f and f_1 have the same value at some point of $\mathfrak{G}$).

extent to which an entire function in general is determined by its zeros. If $G_0(z)$ and $G(z)$ are two entire functions which coincide in the positions and orders of their zeros, then (cf. I, §21, Theorem 4) their quotient is also an entire function, but one with *no* zeros. $G(z)$ and $G_0(z)$ thus differ (cf. statement (B)) by at most a multiplicative entire function with no zeros. Conversely, the presence of such a factor of $G_0(z)$ does not alter the positions or orders of its zeros. In connection with Theorem 1, we express this as follows:

Theorem 2. *Let $G_0(z)$ be a particular entire function. Then, if $h(z)$ denotes an arbitrary entire function,*

$$G(z) = e^{h(z)} \cdot G_0(z)$$

is the most general entire function whose zeros coincide with those of $G_0(z)$ in position and order.

The question of the possibility and method of constructing a particular entire function with arbitrarily prescribed zeros now remains to be settled.

We must begin by restricting our requirements. An entire function has no singularity in the finite part of the plane; therefore, according to I, §21, Theorem 1, it can have only a finite number of zeros in every finite region. The prescribed points consequently must not have a finite limit point. If we make this single restriction, which is in the nature of things, we shall see that an entire function of the kind in question can always be constructed. It will be possible to set it up in the form of a product (analogous to the case of the entire rational functions; cf. (1)) which exhibits the positions and orders of its zeros. We have indeed the following theorem, which is named after its discoverer:

Weierstrass's factor-theorem. *Let any finite or infinite set of points having no finite limit point be prescribed, and associate with each of its points a definite positive integer as order. Then there exists an entire function which has zeros to the prescribed orders at precisely the prescribed points, and is otherwise different from zero. It can be represented as a product (see p. 18 for the final form) from which one can read off again the positions and orders of the zeros. Further, by Theorem 2, if $G_0(z)$ is one such function,*

$$G(z) = e^{h(z)} \cdot G_0(z)$$

is the most general function satisfying the conditions of the problem, if $h(z)$ denotes an arbitrary entire function.[1]

If we regard this fundamental theorem for the moment as having been proved, it follows immediately therefrom, that the first of our two statements concerning the entire rational functions can also be carried over to arbitrary entire functions. For, let $G(z)$ be an arbitrarily given entire function. Then the set of its zeros has no finite limit point. Hence, according to Weierstrass's theorem, another entire function $G_0(z)$, having precisely the same zeros in position and order, can be constructed in the form of a product displaying these. Then, by Theorem 2,

$$G(z) = e^{h_0(z)} \cdot G_0(z),$$

[1] If the entire function to be constructed is to have no zeros, then the factor $G_0(z)$ is to be suppressed, i.e., replaced by unity, which is nevertheless an entire function with the prescribed zeros.

where $h_0(z)$ denotes a suitable entire function. We have thus actually obtained a *factor representation* of the given entire function $G(z)$, from which the positions and orders of its zeros can be read off.

The two statements (A) and (B) concerning entire rational functions have herewith been carried over verbatim to arbitrary entire functions.

The next paragraph is devoted to a proof of Weierstrass's factor-theorem.

Exercises. 1. $\dfrac{\sin iz}{e^{2z} - 1}$ is an entire function with no zeros. (Proof?) Hence, according to Theorem 1, it can be expressed in the form $e^{h(z)}$. How should $h(z)$ be chosen?

2. $\cos iz$ and $e^{2z} + 1$ have the same zeros in position and order. (Proof?) By Theorem 2, the second function can be obtained from the first by multiplying it by a suitable factor of the form $e^{h(z)}$. How should $h(z)$ be chosen?

§2. Proof of Weierstrass's Factor-theorem

As we have already pointed out, the entire function satisfying the conditions of Weierstrass's factor-theorem will be set up in the form of a product; in general, in the form of an infinite product. As with infinite series, we shall assume that the simplest facts in the theory of infinite products with constant factors are familiar to the reader.

Since, however, these are not so universally well-known, and in order to provide a firm foundation for what follows, we present very briefly, without proofs, the

most important definitions and theorems for our purposes.[1]

Definition. *The infinite product*

$$(1) \qquad u_1 \cdot u_2 \cdots u_\nu \cdots = \prod_{\nu=1}^{\infty} u_\nu,$$

*in which the factors are arbitrary complex numbers, is said
to be* **convergent** *(in the stricter sense) if, and only if,
from a certain index on, say for all $\nu > m$, no factor
vanishes, and*

$$\lim_{n \to \infty} (u_{m+1} \cdot u_{m+2} \cdots u_n)$$

*exists and has a finite value distinct from zero. If we call
this limit U_m, then the number*

$$U = u_1 \cdot u_2 \cdots u_m \cdot U_m,$$

which is obviously independent of m, is regarded as the
value *of the infinite product* (1).[2]

---

[1]Detailed proofs are given in K. Knopp, *Theory and Application
of Infinite Series*, translated by R. C. Young, London and
Glasgow, 1928.

[2]With reference to the corresponding definition for infinite
series, one might already be inclined to call the product (1) convergent with the value U if

$$\lim_{n \to \infty} (u_1 \cdot u_2 \cdots u_n) = U.$$

But then *every* product in which only a *single* factor vanishes
would evidently be convergent, and *always* with the same value
zero. Likewise, *every* product such that $|u_\nu| \leq \theta < 1$ for all $\nu > m$
would be convergent, and *always* with the same value zero. To
exclude these cases we employ the more useful definition above,
and, if necessary, draw attention to the restriction it contains
by adding: "in the stricter sense".

The following theorems are easily proved for such convergent infinite products:

Theorem 1. *A convergent product has the value zero if, and only if, one of its factors vanishes.*

Theorem 2. *The infinite product* (1) *is convergent if, and only if, having chosen an arbitrary* $\epsilon > 0$, *an index* n_0 *can be determined such that*

$$| u_{n+1} \cdot u_{n+2} \cdots u_{n+r} - 1 | < \epsilon$$

for all $n > n_0$ *and all* $r \geq 1$ (cf. I, §3, Theorem 4).

Since on the basis of this theorem (let $r = 1$ and $n + 1 = \nu$) it is necessary that $\lim\limits_{\nu \to \infty} u_\nu = 1$, one usually sets the factors of the product equal to $1 + c_\nu$, so that instead of dealing with (1) one is concerned with products of the form

$$(2) \qquad \prod_{\nu=1}^{\infty} (1 + c_\nu).$$

For these, then, $c_\nu \to 0$ is a necessary (*but by no means sufficient*) condition for convergence.

We make use of the following

Definition. *The product* (2) *is said to be absolutely convergent if*

$$\prod_{\nu=1}^{\infty} (1 + | c_\nu |)$$

converges.[1]

[1]The definition which first suggests itself: "Πu_ν shall be called absolutely convergent if $\Pi | u_\nu |$ converges," is not to the purpose, since then *every* convergent product would at the same time converge absolutely.

We then have

Theorem 3. *Absolute convergence is a sufficient condition for ordinary convergence; in other words, the convergence of* $\Pi(1 + |c_\nu|)$ *implies that of* $\Pi(1 + c_\nu)$.

On the basis of this theorem it will be sufficient for our purposes to have convergence criteria for absolutely convergent products. The following two theorems settle completely the question of convergence for these products:

Theorem 4. *The product* $\Pi(1 + \gamma_\nu)$, *with* $\gamma_\nu \geq 0$, *is convergent if, and only if, the series* $\Sigma\gamma_\nu$ *converges.*

Theorem 5. *For* $\Pi(1 + c_\nu)$ *to converge absolutely, it is necessary and sufficient that* Σc_ν *converge absolutely.*[1]
The following theorem is analogous to one on absolutely convergent series:

Theorem 6. *If the order in which the factors of an absolutely convergent product occur is changed in a completely arbitrary manner, the product remains convergent and has the same value.*[2]

---

[1]According to this, $\prod_{\nu=1}^{\infty}(1 - (z^2/\nu^2))$, for example, is absolutely convergent for every value of z, because the series $\Sigma|z^2/\nu^2| = |z|^2 \Sigma(1/\nu^2)$ converges.

[2]In other words, the *commutative law* holds for absolutely convergent infinite products as well as for products having only a finite number of factors. This is not true for non-absolutely convergent products. On the other hand, the *associative law* holds for all convergent products, i.e., one may, in an arbitrary manner, group consecutive factors into one by means of parentheses.

In addition to products with constant factors, we need products whose factors are functions of a complex variable z. We shall write these products in the form

$$(3) \qquad \prod_{\nu=1}^{\infty} (1 + f_\nu(z)).$$

Analogous to the considerations in I, ch. 6, we designate as the *region of convergence* of such a product the set $\mathfrak{M}$ of all those points z which (a) belong to the domain of definition of *every* $f_\nu(z)$, and for which (b) the product (3) is convergent.[1] According to this, the product furnishes a certain value for every z of $\mathfrak{M}$; we say, therefore, that the product represents in $\mathfrak{M}$ a certain (single-valued) function. For our function-theoretical purposes, it is again (cf. I, §19, Theorem 3) particularly important to possess useful conditions under which such a product, in its region of convergence, represents an *analytic function*. The following theorem is adequate:

Theorem 7. *Let $f_1(z), f_2(z), \cdots, f_\nu(z), \cdots$ be an infinite sequence of functions, and suppose a region $\mathfrak{G}$ exists in which all these functions are regular. Let $\sum_{\nu=1}^{\infty} | f_\nu(z) |$ be uniformly convergent in every closed subregion $\mathfrak{G}'$ of $\mathfrak{G}$ (cf. I, p. 74). Then the product (3) is convergent in the entire region $\mathfrak{G}$, and represents a regular function $f(z)$ in $\mathfrak{G}$. Moreover, this function, by Theorem 1, has a zero at those, and only those, points of $\mathfrak{G}$ at which at least one*

---

[1]For instance, the region of convergence of $\prod_{\nu=1}^{\infty} (1 - (z^2/\nu^2))$ is the entire z-plane, according to the last footnote but one.

*of the factors is equal to zero. The order of such a zero is
equal to the sum of the orders to which these factors[1] vanish
there.*

Proof: Let $\mathfrak{G}'$ be an arbitrary closed subregion of $\mathfrak{G}$.
For every $m \geq 0$,

$$\sum_{\nu=m+1}^{\infty} | f_\nu(z) | \text{ , along with } \sum_{\nu=1}^{\infty} | f_\nu(z) | \text{ ,}$$

converges uniformly in $\mathfrak{G}'$. By Theorem 5, the product

$$(4) \qquad \prod_{\nu=m+1}^{\infty} (1 + f_\nu(z))$$

is absolutely convergent in $\mathfrak{G}'$, and represents a certain
function there. Let us call this function $F_m(z)$. Now,
choose the number m such that

$$(5) \quad | f_{n+1}(z) | + | f_{n+2}(z) | + \cdots + | f_{n+r}(z) | < \tfrac{1}{2}$$

for all $n \geq m$, all $r \geq 1$, and all z in $\mathfrak{G}'$ (this is possible by
I, §18). Then $F_m(z)$ is actually regular and distinct from
zero in $\mathfrak{G}'$. Indeed, if, for $n > m$, we set

$$\prod_{\nu=m+1}^{n} (1 + f_\nu(z)) = P_n \text{ and } P_m = 0$$

for abbreviation, we have

$$F_m(z) = \lim_{n \to \infty} P_n$$

$$= \lim_{n \to \infty} [(P_{m+1}-P_m)+(P_{m+2}-P_{m+1})+\cdots+(P_n-P_{n-1})],$$

---

[1]The proof will show that there are only a finite number of
factors in question.

or

(6) $$F_m(z) = \sum_{\nu=m+1}^{\infty} (P_\nu - P_{\nu-1}),$$

and $F_m(z)$ is thus represented by an infinite series. Now the theorems of I, §19 bring us rapidly to our goal. Since, for $n > m$,

$$| P_n | \leq (1 + | f_{m+1}(z) |) \cdots (1 + | f_n(z) |)$$
$$\leq e^{|f_{m+1}(z)|+\cdots+|f_n(z)|} < e^{\frac{1}{2}} < 2,^{1}$$

the inequality

$$| P_\nu - P_{\nu-1} | = | P_{\nu-1} | \cdot | f_\nu(z) | < 2 | f_\nu(z) |$$

is valid for the terms (from the second onward) of the series just obtained. Consequently, the new series (6), along with $\Sigma | f_\nu(z) |$, is uniformly convergent in $\mathfrak{G}'$, and the function $F_m(z)$ defined by that series is a regular function in $\mathfrak{G}'$. It is also distinct from zero there. For, by (5), we have in $\mathfrak{G}'$, for $n \geq m$,

$$| f_{n+1}(z) | < \tfrac{1}{2},$$

and hence, for $\nu \geq m + 1$,

$$| 1 + f_\nu(z) | \geq 1 - | f_\nu(z) | > \tfrac{1}{2},$$

so that no factor of F_m can be equal to zero. Since

$$f(z) = (1 + f_1(z)) \cdots (1 + f_m(z)) \cdot F_m(z),$$

$f(z)$, together with $F_m(z)$, is regular at every point z of $\mathfrak{G}'$, and can vanish at such a point only if one of the

[1]For $x \geq 0,\ 1 + x \leq 1 + x + x^2/2! + \cdots = e^x.$

factors appearing before $F_m(z)$ vanishes. The order of such a zero is then indeed equal to the sum of the orders to which these factors vanish there.

Now let z be an arbitrary point of $\mathfrak{G}$. Since z is *eo ipso* an *interior* point of $\mathfrak{G}$, it is always possible to choose $\mathfrak{G}'$ such that z also belongs to $\mathfrak{G}'$. Hence, the above considerations are valid for the entire region $\mathfrak{G}$, and the proof of the theorem is complete.

Corresponding to the further content of Theorem 3 in I, §19, it is also possible to make an assertion concerning the derivative of $f(z)$. Since the ordinary derivative of a product of many factors is difficult to survey, however, it is more advantageous to choose the so-called **logarithmic derivative**[1] for this purpose.

We then have the following theorem concerning this derivative:

Theorem 8. *Under the hypotheses of Theorem* 7,

$$(7) \qquad \frac{f'(z)}{f(z)} = \sum_{\nu=1}^{\infty} \frac{f'_\nu(z)}{1 + f_\nu(z)}$$

for every point z of $\mathfrak{G}$ at which $f(z) \neq 0$; i.e., the series on the right is convergent for every such z and furnishes the logarithmic derivative of $f(z)$.

Proof: If z is a particular point of the type mentioned, and if the subregion $\mathfrak{G}'$ is chosen so as to contain z, then

---

[1]This is defined as the ordinary derivative divided by the original function. If $F(z) = g_1(z) \cdot g_2(z) \cdots g_k(z)$, and if all the factors are differentiable and distinct from zero at z_0, then

$$\frac{F'(z_0)}{F(z_0)} = \frac{g'_1(z_0)}{g_1(z_0)} + \frac{g'_2(z_0)}{g_2(z_0)} + \cdots + \frac{g'_k(z_0)}{g_k(z_0)}.$$

$$(8) \quad \frac{f'(z)}{f(z)} = \frac{f_1'(z)}{1 + f_1(z)} + \cdots + \frac{f_m'(z)}{1 + f_m(z)} + \frac{F_m'(z)}{F_m(z)}.$$

Since the series (6) converges uniformly in $\mathfrak{G}'$,

$$F_m'(z) = \sum_{\nu=m+1}^{\infty} (P_\nu' - P_{\nu-1}') = \lim_{n \to \infty} P_n',$$

according to I, §19, Theorem 3. Here P_n' denotes the derivative of P_n. Since $F_m(z)$ and all P_n for $n > m$ are not zero,

$$\frac{F_m'(z)}{F_m(z)} = \lim_{n \to \infty} \frac{P_n'}{P_n} = \lim_{n \to \infty} \left(\frac{f_{m+1}'(z)}{1 + f_{m+1}(z)} + \cdots + \frac{f_n'(z)}{1 + f_n(z)} \right)$$

$$= \sum_{\nu=m+1}^{\infty} \frac{f_\nu'(z)}{1 + f_\nu(z)},$$

which, with (8), proves the assertion.

Theorem 9. *The series (7) converges absolutely and uniformly in every closed subregion $\mathfrak{G}''$ of $\mathfrak{G}$ containing no zero of $f(z)$, and hence may be repeatedly differentiated there any number of times term by term.*

Proof: Since none of the factors $(1 + f_\nu(z))$ can vanish in $\mathfrak{G}''$, the absolute value of each remains greater than a positive bound[1], γ_ν say. Since this is certainly greater than $1/2$ for all $\nu > m$ (see above), a positive number γ exists, such that $\gamma_\nu \geq \gamma > 0$ for *all* ν. Then, for all ν and all z in $\mathfrak{G}''$,

$$\left| \frac{f_\nu'(z)}{1 + f_\nu(z)} \right| < \frac{1}{\gamma} \cdot | f_\nu'(z) |.$$

[1]For, $|1 + f_\nu(z)|$, as a continuous function, *attains* its greatest lower bound, and this cannot be zero in $\mathfrak{G}''$.

From the proof of Theorem 3 in I, §19 (cf. also Exercise 2 there) it follows that $\Sigma| f_\nu'(z) |$ converges uniformly in $\mathfrak{G}''$. By the last inequality, this is also true then of the series (7).

Having now familiarized ourselves to some extent with infinite products, it is an easy matter to prove Weierstrass's factor-theorem.

If only a *finite number* of points $z_1, z_2, \cdots, z_k$ with the respective orders $\alpha_1, \alpha_2, \cdots, \alpha_k$ are prescribed, then the product

$$(1) \qquad (z - z_1)^{\alpha_1} (z - z_2)^{\alpha_2} \cdots (z - z_k)^{\alpha_k}$$

is already a solution of the problem, so that this case is settled immediately. If, however, an *infinite number* of points are prescribed as zeros, we cannot proceed quite so simply, because the analogous product would be meaningless in general. This would still be the case if, with regard to the infinite products dealt with, we were to replace (1) by the product

$$(2) \qquad \left(1 - \frac{z}{z_1}\right)^{\alpha_1} \left(1 - \frac{z}{z_2}\right)^{\alpha_2} \cdots \left(1 - \frac{z}{z_k}\right)^{\alpha_k}$$

which serves the same purpose. We therefore proceed somewhat differently—and in this modification lies the originality of Weierstrass's method.

The set of prescribed points is enumerable (see I, p. 10), since every finite region can contain only a finite number of them. They can therefore be arranged in a sequence.[1] The way in which the points are numbered is

[1]This can be done, e.g., by describing circles about the origin with radii 1, 2, 3, $\cdots$, arranging the points as they appear in the consecutive circular rings, and ordering those (only finite in number) which lie in the same ring, according to any rule.

unimportant. However, if the origin, with the order α_0, is contained among them, we shall call this point z_0 and, leaving it aside for the present, arrange the remaining points in an arbitrary, but then fixed, sequence: $z_1, z_2, \cdots, z_\nu, \cdots$. Let the corresponding orders be $\alpha_1, \alpha_2, \cdots, \alpha_\nu, \cdots$. The z_ν are all different from zero; and since they have no finite limit point,

$$z_\nu \to \infty, \; |z_\nu| \to +\infty.$$

Consequently, it is possible (indeed, in many ways) to assign a sequence of positive integers $k_1, k_2, \cdots, k_\nu, \cdots$ such that

$$(3) \qquad \sum_{\nu=1}^{\infty} \alpha_\nu \left(\frac{z}{z_\nu}\right)^{k_\nu}$$

is *absolutely* convergent for *every* z. In fact, it suffices, e.g.[1], to take $k_\nu = \nu + \alpha_\nu$. For, no matter what fixed value z may have, since $z_\nu \to \infty$, we have for all sufficiently large ν

$$\left|\frac{z}{z_\nu}\right| < \tfrac{1}{2}$$

and hence

$$\left| \alpha_\nu \left(\frac{z}{z_\nu}\right)^{\nu+\alpha_\nu} \right| < \alpha_\nu \, (\tfrac{1}{2})^{\nu+\alpha_\nu} < (\tfrac{1}{2})^\nu, \; ^2$$

and the absolute convergence of the series is thus assured.

Let the numbers k_ν be chosen subject to this condi-

[1]Much smaller numbers will often do.
[2]$\alpha/2^\alpha < 1$ for every natural number α.

tion, but otherwise arbitrarily, and keep them fixed. Then we shall prove that the product[1]

$$G_0(z) = z^{\alpha_0} \cdot \prod_{\nu=1}^{\infty} \left[\left(1 - \frac{z}{z_\nu} \right) \cdot \right.$$

$$\left. \cdot \exp \left\{ \frac{z}{z_\nu} + \frac{1}{2} \left(\frac{z}{z_\nu} \right)^2 + \cdots + \frac{1}{k_\nu - 1} \left(\frac{z}{z_\nu} \right)^{k_\nu - 1} \right\} \right]^{\alpha_\nu}$$

(Weierstrass's factor-theorem)

represents an entire function with the required properties[2]. (Here the factor z^{α_0} appearing before the product symbol is to be suppressed in case the origin is not one of the prescribed zeros (see above). Likewise, if one of the numbers k_ν is equal to unity, the corresponding exponential factor simply does not appear.)

The proof of this assertion is now very simple. To be able to apply our theorems on products, we set the factors of our infinite product equal to $1 + f_\nu(z)$. According to Theorem 7, we must then merely prove that

$$(4) \quad \sum_{\nu=1}^{\infty} | f_\nu(z) | = \sum_{\nu=1}^{\infty} \left| \left[\left(1 - \frac{z}{z_\nu} \right) \right. \right.$$

$$\left. \left. \exp \left\{ \frac{z}{z_\nu} + \cdots + \frac{1}{k_\nu - 1} \left(\frac{z}{z_\nu} \right)^{k_\nu - 1} \right\} \right]^{\alpha_\nu} - 1 \right|$$

[1]We shall find it convenient sometimes to write exp z instead of e^z.

[2]The exponentials in the brackets ensure the convergence of the product, which would, in general, diverge without these. They are therefore called the *convergence-producing factors*.

Weierstrass called the expressions in the brackets *primary factors*.

converges uniformly in *every* bounded region. For then
the entire plane can be taken as the region $\mathfrak{G}$ of Theorem 7, according to which the infinite product, and
consequently also $G_0(z)$, is an entire function. On account of the form of the factors of $G_0(z)$, the second part
of Theorem 7 at once yields that $G_0(z)$ also possesses the
required properties. The uniform convergence of the
series (4) in the circle about the origin with radius
R ($R > 0$ arbitrary, but fixed) is established as follows:

Since the series (3) also converges for $z = R$, and since
$z_\nu \to \infty$, m can be chosen so large that

$$(5) \qquad \alpha_\nu \left| \frac{R}{z_\nu} \right|^{k_\nu} < \tfrac{1}{2} \quad \text{and} \quad \frac{R}{|z_\nu|} < \tfrac{1}{2}$$

for all $\nu > m$. Let us for the moment replace z/z_ν by
u, k_ν by k, and α_ν by α. Then, for $\nu > m$, the νth term
of the series (4) has the form

$$\left| \left[(1 - u) \exp\left\{ u + \frac{u^2}{2} + \cdots + \frac{u^{k-1}}{k-1} \right\} \right]^\alpha - 1 \right|$$

with $\begin{cases} |u| < \tfrac{1}{2} \quad \text{and} \\ \alpha |u|^k < \tfrac{1}{2}. \end{cases}$

Now for $|u| < 1$ we can set[1]

$$1 - u = \exp\left\{ -u - \frac{u^2}{2} - \frac{u^3}{3} - \cdots \right\},$$

[1] For, the series in the exponent on the right has for its sum
the principal value of $\log (1 - u)$.

According to the theorem formulated on p. 4, footnote, this
simple fact also follows from the coincidence of both sides for
$u = 0$ and the equality of their logarithmic derivatives for all
$|u| < 1$.

so that this νth term is further equal to

$$\left| \exp\left\{ \alpha\left(-\frac{u^k}{k} - \frac{u^{k+1}}{k+1} - \cdots \right) \right\} - 1 \right|,$$

and hence[1]

$$\leq \exp\left\{ \alpha\left(\frac{|u|^k}{k} + \frac{|u|^{k+1}}{k+1} + \cdots \right) \right\} - 1$$

$$\leq \exp\left\{ \alpha |u|^k (1 + |u| + |u|^2 + \cdots) \right\} - 1$$

$$< e^{2\alpha|u|^k} - 1,$$

because $|u| < \frac{1}{2}$. Further, since $e^x - 1 \leq xe^x$ for $x \geq 0$,[2] the νth term is less than or equal to

$$2\alpha |u|^k e^{2\alpha|u|^k} < 6\alpha|u|^k,$$

the exponent of e being smaller than one, according to (5). Hence, for all sufficiently large ν and all $|z| \leq R$ we have

$$|f_\nu(z)| < 6\alpha_\nu \left| \frac{z}{z_\nu} \right|^{k_\nu} \leq 6\alpha_\nu \left| \frac{R}{z_\nu} \right|^{k_\nu}.$$

But these are positive numbers whose sum converges (because of the manner in which the k_ν were chosen). Therefore, by Weierstrass's M-test, I, §18, $\Sigma |f_\nu(z)|$ is uniformly convergent in the circle with radius R about the origin as center, and so the proof of the Weierstrass factor-theorem is complete.

---

[1] $|e^w - 1| \leq |w| + |w^2/2!| + \cdots = e^{|w|} - 1$ for every complex number w.

[2] $e^x - 1 = x + x^2/2! + \cdots = x(1 + x/2! + x^2/3! + \cdots) \leq xe^x$.

Exercises. 1. Prove Theorems 1-6.

2. Establish the convergence of, and evaluate, each of the following products with constant factors:

a) $\prod\limits_{n=1}^{\infty} \left(1 + \dfrac{1}{n(n+2)}\right)$;   b) $\prod\limits_{n=2}^{\infty} \left(1 - \dfrac{2}{n(n+1)}\right)$;

c) $\prod\limits_{n=2}^{\infty} \left(\dfrac{n^3-1}{n^3+1}\right)$.

3. Determine the region of convergence of each of the following products:

a) $\prod\limits_{n=1}^{\infty} (1 - z^n)$;   b) $\prod\limits_{n=0}^{\infty} (1 + z^{2^n})$;   c) $\prod\limits_{n=2}^{\infty} \left(1 - \dfrac{1}{n^z}\right)$;

d) $\prod\limits_{p} \left(1 - \dfrac{1}{p^s}\right)$, if p runs over all the prime numbers;

e) $\prod\limits_{n=1}^{\infty} (1 + c_n z)$, if Σc_n is an absolutely convergent series.

4. Prove the following formulas:

a) $\prod\limits_{p} \dfrac{1}{1 - p^{-z}} = \sum\limits_{n=1}^{\infty} \dfrac{1}{n^z}$ (cf. 3 c and d);

b) $\prod (1 + z^{2^n}) = \dfrac{1}{1 - z}$ (cf. 3 b).

5. What values have the coefficients μ_n on the right-hand side in the equation

$$\prod\limits_{p} \left(1 - \dfrac{1}{p^z}\right) = \sum\limits_{n=1}^{\infty} \dfrac{\mu_n}{n^z},$$

in which p again is to run over all the prime numbers.

6. If $z_1, z_2, \cdots, z_n, \cdots$ is any sequence of numbers which tends to ∞, then, if all $z_n \neq 0$,

$$\sum_{n=1}^{\infty} \left(\frac{z}{z_n}\right)^{\log n}$$

is convergent for every z. Hence, what smaller numbers k_ν can one always choose in the proof of Weierstrass's factor-theorem instead of the ones used in the text?

7. Prove the following transfer of Weierstrass's factor-theorem to the region of the unit circle:

Let $z_1, z_2, \cdots, z_n, \cdots$ be an arbitrary sequence of distinct points inside the unit circle, which have no limit point in the *interior* of this circle (but only on its circumference). Let $\alpha_1, \alpha_2, \cdots, \alpha_n, \cdots$ be a sequence of arbitrary positive integers. Then it is always possible to construct a function $f(z)$ (and indeed, in a form closely analogous to the Weierstrass product) which is regular in the unit circle, and there has zeros of the orders α_n at precisely the points z_n, respectively, (and no others).

8. With the aid of the preceding theorem, construct functions which have the unit circle for a natural boundary.

§3. Examples of Weierstrass's Factor-theorem

Since the formation of entire functions with prescribed zeros is extremely simple—it was only somewhat more laborious to carry out carefully all the proofs, —one can easily construct any number of examples.

The product is simplest if the prescribed zeros and orders are such, that the series $\sum \dfrac{\alpha_\nu}{z_\nu}$, and consequently, for every z, the series $\sum \alpha_\nu \left(\dfrac{z}{z_\nu}\right)$, converges absolutely for our sequence $z_1, z_2, \cdots$. For then it is possible to take all $k_\nu = 1$, and the desired function is obtained simply in the form

$$G_0(z) = z^{\alpha_0} \cdot \prod_{\nu=1}^{\infty} \left(1 - \frac{z}{z_\nu}\right)^{\alpha_\nu}.$$

If, e.g., the points $0, 1, 4, 9, \cdots, \nu^2, \cdots$ are to be zeros of order unity, then

$$G(z) = e^{h(z)} \cdot z \cdot \prod_{\nu=1}^{\infty} \left(1 - \frac{z}{\nu^2}\right),$$

with $h(z)$ an arbitrary entire function, is the most general solution of the problem. If the points $1, 8, \cdots$, $\nu^3, \cdots$ are to be zeros of respective orders $1, 2, \cdots$, $\nu, \cdots$, then

$$G(z) = e^{h(z)} \cdot \prod_{\nu=1}^{\infty} \left(1 - \frac{z}{\nu^3}\right)^{\nu}$$

is the most general solution.

In addition to these simple examples, whose number is easily enlarged, we shall now present several applications of the factor theorem which are of particular function-theoretical importance.

1st Example: sin πz. Consider the problem of constructing an entire function which has zeros, of order unity, at precisely all the real lattice points (i.e., at 0, $\pm 1, \pm 2, \cdots$). We number these points so that $z_0 = 0$, $z_1 = +1, z_2 = -1, \cdots$, $z_{2\nu-1} = \nu, z_{2\nu} = -\nu, \cdots$,

$(\nu = 1, 2, \cdots)$. The series

$$\sum_{\nu=1}^{\infty} \left(\frac{z}{z_\nu}\right)^2 = z^2 \cdot \sum_{\nu=1}^{\infty} \frac{1}{z_\nu^2}$$

is absolutely convergent for every z, and we can therefore take all $k_\nu = 2$. Then

$$G(z) = e^{h(z)} \cdot z \cdot \prod_{\nu=1}^{\infty} \left[\left(1 - \frac{z}{z_\nu}\right) e^{z/z_\nu} \right]$$

$$= e^{h(z)} \cdot z \cdot \prod_{\nu=1}^{\infty} \left[\left(1 - \frac{z}{\nu}\right) e^{z/\nu} \right] \left[\left(1 + \frac{z}{\nu}\right) e^{-z/\nu} \right]$$

$$= e^{h(z)} \cdot z \cdot \prod_{\nu=1}^{\infty} \left(1 - \frac{z^2}{\nu^2}\right) \quad^1$$

is the most general solution of the problem.

Since the function $\sin \pi z$ is evidently also a solution of the problem, it must be contained in the expression just found. That is, there exists a certain entire function, which we shall call $h_0(z)$, such that

$$(1) \qquad \sin \pi z = e^{h_0(z)} \cdot z \cdot \prod_{\nu=1}^{\infty} \left(1 - \frac{z^2}{\nu^2}\right).$$

If we succeed in obtaining this function $h_0(z)$, we shall have the *factor representation* of $\sin \pi z$ in the sense of §1 (see p. 6).

The function $h_0(z)$ certainly can not be ascertained from a knowledge of the zeros *alone*. On the contrary, for its determination we must make use of further properties of the particular function $\sin \pi z$; e.g., its power-

[1]The transformations are justified according to p. 10, footnote 2.

series expansion, its periodicity properties, the conformal map effected by it, its behavior at infinity, etc.
We sketch briefly a method for determining $h_0(z)$.[1]

First, we show that $h_0''(z)$ is a constant. According to
§2, Theorem 8, it follows from (1) that

$$(2) \quad \pi \cot \pi z = h_0'(z) + \frac{1}{z} + \sum_{\nu=1}^{\infty} \left(\frac{1}{z - \nu} + \frac{1}{z + \nu} \right).$$

According to Theorem 9, this expression may be differentiated repeatedly term by term. Thus,

$$-\frac{\pi^2}{\sin^2 \pi z} = h_0''(z) - \frac{1}{z^2} - \sum_{\nu=1}^{\infty} \left(\frac{1}{(z - \nu)^2} + \frac{1}{(z + \nu)^2} \right),$$

or, written more briefly,

$$h_0''(z) = \sum_{\nu=-\infty}^{+\infty} \frac{1}{(z - \nu)^2} - \frac{\pi^2}{\sin^2 \pi z}.$$

This relation holds in every closed region which contains no real lattice points. If we replace z by $z + 1$ in
the right-hand member, it is not altered; because
$\sin^2 \pi z$ has the period $+1$, and

$$\sum_{\nu=-\infty}^{+\infty} \frac{1}{(z + 1 - \nu)^2} = \sum_{\nu=-\infty}^{+\infty} \frac{1}{(z - (\nu - 1))^2} = \sum_{\mu=-\infty}^{+\infty} \frac{1}{(z - \mu)^2}.$$

[1]Here we are concerned with a *typical question*: One has two
analytical expressions: $A_1(z)$ and $A_2(z)$, say, as in the present
case the already familiar power-series representation of $\sin \pi z$ on
the one hand, and the infinite product $z \cdot \Pi(1 - (z^2/\nu^2))$ on the other.
In the course of an investigation, one is led to the conjecture that
both expressions represent the same function, or stand in some
simple relationship to one another. How can this be proved?
The determination of $h_0(z)$ carried out in the text shows that
even in the present apparently very simple instance such an
identification is not very easily made.

Hence, $h_0''(z)$ is an entire function with the period $+1$. In order to show that $h_0''(z)$ is a constant, it is sufficient, by I, §28, 1, to show that $|\, h_0''(z)\, |$ cannot become arbitrarily large. On account of the periodicity of $h_0''(z)$ which we just established, it is sufficient, for this purpose, to show that a constant K exists such that $|\, h_0''(z)\, | < K$ for all $z = x + iy$ for which $0 \leq x \leq 1$ and $|\, y\, | \geq 1$.

Now for these z,

$$\left|\sum_{\nu=-\infty}^{+\infty}\frac{1}{(z-\nu)^2}\right| \leq \sum_{\nu=-\infty}^{+\infty}\frac{1}{(x-\nu)^2+y^2} \leq 2\sum_{n=0}^{\infty}\frac{1}{n^2+y^2};$$

and, since $\sin \pi z = (1/2i)(e^{i\pi z} - e^{-i\pi z})$,

$$\left|\frac{\pi^2}{\sin^2 \pi z}\right| = \frac{4\pi^2}{e^{2\pi y}+e^{-2\pi y}-2\cos 2\pi x} < \frac{4\pi^2}{e^{2\pi|y|}-2}$$

for those z. Consequently,

$$\left|\, h_0''(z)\, \right| < 2\sum_{n=0}^{\infty}\frac{1}{n^2+y^2}+\frac{4\pi^2}{e^{2\pi|y|}-2}$$

there, and this expression certainly remains less than a fixed bound for all $|\, y\, | \geq 1$. Hence,

$$h_0''(z) = \text{constant} = c''.$$

According to the inequality just obtained, $|\, h_0''(z)\, |$ is arbitrarily small if $|\, y\, |$ is sufficiently large; hence c'' must be equal to zero. Therefore

$$h_0''(z) = 0, \qquad h_0'(z) = \text{constant} = c',$$

and hence by (2)

$$\pi \cot \pi z = c' + \frac{1}{z} + \sum_{\nu=1}^{\infty}\frac{2z}{z^2-\nu^2}.$$

If we substitute $-z$ for z in this equality, we see that $c' = -c'$, and hence $c' = 0$. Then $h_0(z)$ and $e^{h_0(z)}$ are also constant. Therefore

$$\sin \pi z = c \cdot z \cdot \prod_{\nu=1}^{\infty} \left(1 - \frac{z^2}{\nu^2}\right).$$

If we divide through by z and allow z to approach zero, we obtain $\pi = c$. Thus,

$$\sin \pi z = \pi z \cdot \prod_{\nu=1}^{\infty} \left(1 - \frac{z^2}{\nu^2}\right),$$

valid for *all z*, is the *product representation of the sine-function* which we set out to find.

2d Example: Weierstrass's σ-function. Let ω and ω' be two non-zero numbers whose ratio is not real (or: which do not lie in a straight line with the origin). Then an entire function is to be constructed having zeros, of order unity, at all points of the form

$$k\omega + k'\omega', \qquad \begin{cases} k = 0, \pm 1, \pm 2, \cdots \\ k' = 0, \pm 1, \pm 2, \cdots, \end{cases}$$

and at no other points.

Draw the straight lines L, L' joining the origin to the points ω, ω', respectively (see Fig. 1). Mark the points $k\omega$ on L and $k'\omega'$ on L', and through each of these points draw a line parallel to L', L, respectively. The points of intersection of these two families of parallel lines are precisely the given points $k\omega + k'\omega'$. They are the *"lattice points of a network of parallelograms"* determined by ω and ω'.

We can enumerate these lattice points in the follow-

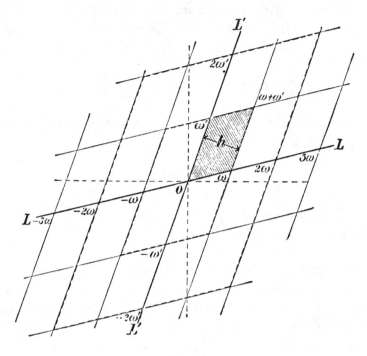

Fig. 1.

ing manner. Consider the parallelograms whose centers lie at the origin, and whose sides are parallel to L, L' and have in turn the lengths 2ω, 4ω, 6ω, 8ω, $\cdots$, $2\omega'$, $4\omega'$, $6\omega'$, $8\omega'$, $\cdots$, respectively. These sides are indicated by interrupted lines in Fig. 1. Now, start with the point O, and number the points lying on the sides of the successive parallelograms; beginning on each parallelogram with the point $k\omega$, and traversing the parallelogram in the mathematically positive sense. We thus obtain a sequence of lattice points, which begins with

$$0, \omega, \omega + \omega', \omega', -\omega + \omega', -\omega, -\omega - \omega', -\omega',$$

$$\omega - \omega', 2\omega, 2\omega + \omega', 2\omega + 2\omega', \cdots .$$

Keeping the points in this order, we denote them by $z_0, z_1, z_2, \cdots$. First we shall show that the series

$$\sum_{\nu=1}^{\infty} \left(\frac{z}{z_\nu}\right)^3$$

is absolutely convergent for every z. Let us number the parallelograms, along which we just counted the lattice points, the 1st, 2d, 3d, $\cdots$ according to size. On the pth one of these lie precisely $8p$ of our lattice points, whose absolute values, moreover, are greater than or equal to ph, if h denotes the smaller of the two altitudes of the *"fundamental parallelogram"* with vertices 0, $\omega, \omega + \omega', \omega'$. Hence, the points of the pth parallelogram contribute to the series $\sum \left|\frac{z}{z_\nu}\right|^3$ an absolute value which is less than or equal to

$$8p \left(\frac{|z|}{ph}\right)^3 = \frac{8 |z|^3}{h^3} \cdot \frac{1}{p^2}.$$

Since $\Sigma (1/p^2)$ converges, the series above also converges absolutely for every z. It is therefore sufficient to take all $k_\nu = 3$ in the Weierstrass product, and

$$G_0(z) = z \cdot \prod_{\nu=1}^{\infty} \cdot \left[\left(1 - \frac{z}{z_\nu}\right) \exp\left\{\frac{z}{z_\nu} + \frac{1}{2}\left(\frac{z}{z_\nu}\right)^2\right\}\right]$$

is, with the meaning given to the z_ν, an entire function with the required properties. In the Weierstrassian

theory of elliptic functions, this function is called the
Sigma-function belonging to the pair of periods (ω, ω'),
and is denoted by

$$\sigma(z) = \sigma(z \mid \tfrac{1}{2}\omega, \tfrac{1}{2}\omega').$$

Because of the absolute convergence of the product, the
order in which the factors appear does not matter any
more (see §2, Theorem 6). Hence, without further
establishing the sequence of the lattice points, we can
write

$$\sigma(z \mid \tfrac{1}{2}\omega, \tfrac{1}{2}\omega')$$

$$= z \cdot \prod_{k,k'}{}' \left[\left(1 - \frac{z}{k\omega + k'\omega'} \right) \cdot \right.$$

$$\left. \cdot \exp \left\{ \frac{z}{k\omega + k'\omega'} + \frac{1}{2} \left(\frac{z}{k\omega + k'\omega'} \right)^2 \right\} \right].$$

Here k and k' take on independently of each other all
positive and negative integral values and zero, without,
however, being zero *simultaneously*. This last restriction
is indicated by the accent after the product symbol.

3d Example. Finally, we shall construct an entire
function which has zeros, of order unity, at $z_0 = 0$,
$z_1 = -1, z_2 = -2, \cdots, z_\nu = -\nu, \cdots$, and at no other
point. Here it is obviously sufficient again to take all
$k_\nu = 2$, so that

$$G(z) = e^{h(z)} \cdot z \cdot \prod_{\nu=1}^{\infty} \left[\left(1 + \frac{z}{\nu} \right) e^{-z/\nu} \right]$$

is the most general function with the required properties. It is closely related to the so-called (**Eulerian**) **Gamma-function,** which, for real values of the argument, is familiar to the reader from the integral calculus, and which was defined by Gauss for arbitrary complex $z \neq 0, -1, -2, \cdots$ by means of the limit

$$\Gamma(z) = \lim_{n \to \infty} \frac{n!\, n^z}{z(z+1)(z+2) \cdots (z+n)}$$

$(n^z = e^{z \log n}$, with $\log n$ real and positive).[1]

It is easy to see that this limit exists for all z in question. For if we write the reciprocal of the expression under consideration in the form

$$e^{-z \log n} \cdot z \cdot \left(1 + \frac{z}{1}\right)\left(1 + \frac{z}{2}\right) \cdots \left(1 + \frac{z}{n}\right),$$

[1]After the elementary functions, the Gamma-function is one of the most important functions of analysis. It is met with in the most varied investigations of pure and applied mathematics, from the theory of numbers to theoretical physics, so that an intimate knowledge of its analytical properties is absolutely indispensable.

The first study of this function is to be traced back to the problem of *interpolating* the sequence of factorials $0!(=1), 1!, 2!, \cdots$; i.e., joining the points $(\nu, \nu!)$ or (as one usually writes, following Euler) the points with coordinates $x = \nu + 1, y = \nu! \ (\nu = 0, 1, 2, \cdots)$ by as simple a curve as possible. This is the problem of finding the simplest real function $y = F(x)$ of the real variable x, such that $y = \nu!$ for $x = \nu + 1$. Euler gave as a solution the integral $\int_0^\infty e^{-t} t^{z-1}\, dt$, which converges for all $\Re(z) > 0$; Gauss, the limit mentioned in the text. Both solutions yield the *same* function for $\Re(z) > 0$. For lack of space we must suppress the proof of this last assertion. An especially elementary proof has been given by A. Pringsheim, Math. Ann. 31 (1888), pp. 455-481.

it is further equal to

$$\left[\exp \left\{ \left(1 + \frac{1}{2} + \cdots + \frac{1}{n} - \log n \right) z \right\} \right] \cdot$$

$$\cdot z \cdot \prod_{\nu=1}^{n} \left[\left(1 + \frac{z}{\nu} \right) e^{-z/\nu} \right].$$

Now, as is well known,

$$\lim_{n \to \infty} \left[\left(1 + \frac{1}{2} + \cdots + \frac{1}{n} \right) - \log n \right] = C$$

exists.[1] Therefore, as $n \to \infty$, our last expression tends

[1] It is immediately evident from the geometrical interpretation of the integral as a plane area, that

$$1/n > \int_{n}^{n+1} (dx/x) > 1/(n+1) \text{ and hence}$$

$$(1/n) - (1/(n+1)) > (1/n) - \int_{n}^{n+1} (dx/x) > 0$$

for every $n = 1, 2, \cdots$. If we set $(1/n) - \int_{n}^{n+1} (dx/x) = \gamma_n$, then

$0 < \gamma_n < (1/n) - (1/(n+1))$. Therefore $\Sigma \gamma_n = C$ is convergent, with $0 < C < 1$. Consequently

$$(\gamma_1 + \gamma_2 + \cdots + \gamma_{n-1})$$

$$= [1 + (1/2) + \cdots + (1/(n-1))] - \int_{1}^{n} (dx/x) \to C;$$

and, since $1/n \to 0$ and $\int_{1}^{n} (dx/x) = \log n,$

$$[(1 + (1/2) + \cdots + (1/n)) - \log n] \to C.$$

C is called Euler's or Mascheroni's constant. Its value lies between zero and one; more precisely: $C = 0.5772156649 \cdots$.

(actually for every z) to the value of the entire function

$$K(z) = e^{Cz} \cdot z \cdot \prod_{\nu=1}^{\infty} \left[\left(1 + \frac{z}{\nu} \right) e^{-z/\nu} \right],$$

which results from the solution of our last example by setting $h(z) = Cz$. Since $K(z)$ is certainly different from zero for $z \neq 0, -1, -2, \cdots$, the Gaussian limit as formulated above exists and is equal to $\dfrac{1}{K(z)}$. It thus defines a single-valued analytic function, namely, the reciprocal of the entire function $K(z)$. (For further details, see §6, 3d Example).

Exercises. 1. Derive the values of the following three products from the sine-product:

a) $$\frac{2}{1} \cdot \frac{2}{3} \cdot \frac{4}{3} \cdot \frac{4}{5} \cdot \frac{6}{5} \cdot \frac{6}{7} \cdots$$

(Wallis's Product)

b) $$\frac{2}{1} \cdot \frac{2}{3} \cdot \frac{6}{5} \cdot \frac{6}{7} \cdot \frac{10}{9} \cdot \frac{10}{11} \cdot \frac{14}{13} \cdots$$

c) $$2 \cdot \frac{2}{3} \cdot \frac{4}{3} \cdot \frac{8}{9} \cdot \frac{10}{9} \cdot \frac{14}{15} \cdot \frac{16}{15} \cdots .$$

2. Obtain the product expansions of the following entire functions:

a) $e^z - 1$; b) $e^z - e^{z_0}$; c) $\sin z - \sin z_0$; $\cos z - \cos z_0$.

3. Demonstrate the existence of entire functions which assume arbitrarily prescribed values $w_1, w_2, \cdots, w_n, \cdots$ at arbitrarily assigned points $z_1, z_2, \cdots, z_n, \cdots$, respectively, having no finite limit point.

CHAPTER 2

MEROMORPHIC FUNCTIONS

§4. Mittag-Leffler's Partial-fractions-theorem

The fractional rational functions are completely characterized in a purely function-theoretical manner by Theorems 1-3 in I, §35. Analogous to our procedure in the preceding chapter, we express the fundamental properties of these functions in the following two statements:

(A) For every (fractional) rational function there is a so-called *decomposition into partial fractions*, which displays its poles and the corresponding principal parts.

Thus, let $f_0(z)$ be the given rational function, and let $z_1, z_2, \cdots, z_k$ be its poles with the corresponding principal parts

$$(1)\quad h_\nu(z) = \frac{a_{-1}^{(\nu)}}{z - z_\nu} + \frac{a_{-2}^{(\nu)}}{(z - z_\nu)^2} + \cdots + \frac{a_{-\alpha_\nu}^{(\nu)}}{(z - z_\nu)^{\alpha_\nu}},$$

$$(\nu = 1, 2, \cdots, k).$$

Then we can set

$$(2)\quad f_0(z) = g_0(z) + h_1(z) + h_2(z) + \cdots + h_k(z),$$

where $g_0(z)$ is a suitable *entire* rational function. We infer at once from this decomposition into partial fractions, that every other rational function $f(z)$ having the same poles with the same respective principal parts can differ from $f_0(z)$ in the term $g_0(z)$ alone. Furthermore, one can arbitrarily assign these poles and their principal parts. In other words:

(B) It is always possible to construct a rational function whose poles and their principal parts are prescribed. This function can be represented as a partial-fractions decomposition which displays these poles and their principal parts. The most general function of this kind is obtained from a particular one by adding to it an arbitrary *entire* rational function.

These fundamental facts concerning rational functions can again be carried over in all particulars to the more general class of meromorphic functions, whose definition we have already indicated in the Introduction, and which we shall now state more precisely.

Definition. *A single-valued function shall—without regard to its behavior at infinity—be called meromorphic, if it has no singularities other than (at most) poles in the entire plane.*

On the basis of this definition we have the following theorem:

Theorem 1. *A meromorphic function has in every finite region at most a finite number of poles.*

For otherwise there would exist a finite limit point of poles, and this point would be singular, but certainly not a pole.

According to this, the rational functions are special cases of meromorphic functions, and the entire functions must also be regarded as such.

The function $\dfrac{1}{\sin z}$ is meromorphic; because in the finite part of the plane it has a singularity, namely a pole of order unity, only wherever sin z has a zero. We see, likewise, that $\cot z = \dfrac{1}{\sin z} \cdot \cos z$ and $\tan z$ are meromorphic functions. More generally, if $G(z)$ denotes any entire function, its reciprocal, $1/G(z)$, is a meromorphic function (and hence, e.g., the function $\Gamma(z) = 1/K(z)$, considered at the end of the preceding paragraph, is meromorphic). For, $1/G(z)$ has poles (but otherwise no singularities) at those, and only those, points at which $G(z)$ has zeros; and the orders of both are the same. If $G_1(z)$ is an entire function which has no zeros in common with $G(z)$, we see that $G_1(z)/G(z)$ is a meromorphic function whose poles coincide in position and order (although, in general, not in their principal parts) with those of $1/G(z)$.[1]

We now again set ourselves the **problem** which corresponds to the second statement (B). We propose to investigate whether, and how, one can construct a meromorphic function if its poles and the corresponding principal parts are prescribed, and to what extent a meromorphic function is determined by these conditions.

This last question can be answered immediately. If $M_0(z)$ and $M(z)$ are two meromorphic functions which

---

[1]This last example represents the most general case; for there is the following **Theorem:** *Every meromorphic function $f(z)$ can be expressed as the quotient of two entire functions having no zeros in common.* Proof: The poles of $f(z)$ have no finite limit point. According to Weierstrass's factor-theorem, we can construct an entire function $G(z)$ whose *zeros* coincide in position and order with the *poles* of $f(z)$. Then $f(z) \cdot G(z)$ is evidently an *entire* function, $G_1(z)$. Hence, $f(z) = G_1(z)/G(z)$; and $G_1(z)$ has no zero in common with $G(z)$.

coincide in their poles and the corresponding principal parts, then their difference, $M(z) - M_0(z)$, is evidently an *entire* function. Consequently, they differ by at most an (additive) entire function ("a meromorphic function with no poles"). Conversely, since the addition of such a function to $M_0(z)$ does not alter its poles or the corresponding principal parts, we are able to say the following:

Theorem 2. *Let $M_0(z)$ be a particular meromorphic function. Then, if $G(z)$ denotes an arbitrary entire function,*

$$M(z) = M_0(z) + G(z)$$

is the most general meromorphic function which coincides with $M_0(z)$ in its poles and the corresponding principal parts.

There remains only the investigation of the possibility and method of constructing a particular meromorphic function with arbitrarily prescribed poles.

According to Theorem 1, the set of assigned poles cannot have a finite limit point. If this is excluded, however, then the problem posed can be solved without any further restriction. The following theorem is named after its discoverer:

Mittag-Leffler's partial-fractions-theorem. *Let any finite or infinite set of points having no finite limit point be prescribed, and associate with each of its points a principal part, i.e., a rational function of the special form (1). Then there exists a meromorphic function which has poles with the prescribed principal parts at precisely the prescribed points, and is otherwise regular. It can be* *represented in the form of a partial-fractions decomposition (see p. 40 for the final form) from which one can read off again the poles along with their principal parts. Further, by Theorem 2, if $M_0(z)$ is one such function,*

$$M(z) = M_0(z) + G(z)$$

is the most general function satisfying the conditions of the problem, if $G(z)$ denotes an arbitrary entire function.

This theorem solves the problem which corresponds to statement (B) concerning rational functions. Let us regard it for the moment as having been proved. Then the solution of the problem corresponding to statement (A) also follows immediately therefrom. For, let $M(z)$ be an arbitrarily given meromorphic function. The set of its poles has no finite limit point. Hence, according to Mittag-Leffler's theorem, another meromorphic function, $M_0(z)$, having the same poles and principal parts as $M(z)$, can be constructed in the form of a partial-fractions decomposition displaying these. Then, by Theorem 2,

$$M(z) = M_0(z) + G_0(z),$$

where $G_0(z)$ denotes a suitable entire function. We have thus actually obtained a decomposition of the *given* meromorphic function $M(z)$ into partial fractions, from which its poles and the corresponding principal parts can be read off.

Exercises. 1. cot z and $\dfrac{2i}{e^{2iz} - 1}$ are two meromorphic

functions which coincide in their poles and the corresponding principal parts. (Proof?) According to Theorem 2, the first differs from the second only by an (additive) entire function. Find this function.

2. The same for the functions

$$\frac{1}{2i \cos z} \text{ and } \frac{\sin z}{e^{2iz} + 1}.$$

§5. Proof of Mittag-Leffler's Theorem

If the function to be constructed is to have no poles at all, then every entire function is a solution of the problem. If it is to have the finitely many poles z_1, $z_2, \cdots, z_k$ with the respective principal parts $h_1(z)$, $h_2(z), \cdots, h_k(z)$, then evidently

$$M_0(z) = h_1(z) + h_2(z) + \cdots + h_k(z)$$

is a solution. If, however, an infinite number of poles are prescribed, we cannot attain our goal so simply; because the analogous series, now infinite, would diverge in general. Nevertheless, we can *produce* the convergence, as in §2, by means of a suitable modification of the terms of the series.·

To this end, we make exactly the same agreements regarding the enumeration of the poles as we did in §2, p. 17 in connection with the zeros. If the origin is a prescribed pole, we denote it by z_0 and leave it aside for the time being. Let $h_0(z)$, $h_1(z)$, $\cdots, h_\nu(z)$, $\cdots$ be the principal parts corresponding to the points $z_0, z_1, \cdots$, $z_\nu, \cdots$; $h_\nu(z)$ is understood to be an expression of the type appearing in formula (1), p. 34. Each of these functions $h_\nu(z)$, $\nu = 1, 2, 3, \cdots$, is regular in a neighborhood of the origin. Its power-series expansion

$$h_\nu(z) = a_0^{(\nu)} + a_1^{(\nu)}z + a_2^{(\nu)}z^2 + \cdots \qquad (\nu = 1, 2, \cdots)$$

for this neighborhood converges for all $|z| < |z_\nu|$; hence, it is *uniformly* convergent for all $|z| \leq \frac{1}{2}|z_\nu|$. Consequently (for every $\nu = 1, 2, 3, \cdots$) an integer n_ν can be determined such that the remainder of the power series after the n_νth term remains, in absolute value, less than any preassigned positive number, e.g., $\frac{1}{2^\nu}$. We denote the sum of the first n_ν terms of the series by $g_\nu(z)$. Thus, $g_\nu(z)$ is an entire rational function of degree n_ν:

$$g_\nu(z) = a_0^{(\nu)} + a_1^{(\nu)}z + \cdots + a_{n_\nu}^{(\nu)}z^{n_\nu} \quad (\nu = 1, 2, 3, \cdots),$$

and for all $|z| \leq \frac{1}{2}|z_\nu|$ we have

$$|h_\nu(z) - g_\nu(z)| < \frac{1}{2^\nu}.$$

Then

$$M_0(z) = h_0(z) + \sum_{\nu=1}^{\infty} [h_\nu(z) - g_\nu(z)]$$

(Mittag-Leffler's partial-fractions-theorem)

is a meromorphic function satisfying the conditions of the theorem. (If the origin is not assigned as a pole, the term $h_0(z)$ must, of course, be omitted.)

To prove this, we must merely show that the right-hand side defines an analytic function having in every finite domain, e.g., a circle with radius R about the origin as center, exactly the prescribed singularities and no others.

Now, $|z_\nu| \to +\infty$. Therefore it is possible to choose m so large, that $|z_\nu| > 2R$, and hence $R < \frac{1}{2}|z_\nu|$, for all $\nu > m$. Then, for all $|z| \leq R$ and all $\nu > m$,

$|z| < \frac{1}{2}|z_\nu|$ and consequently $|h_\nu(z) - g_\nu(z)| < \dfrac{1}{2^\nu}$.

Hence, for all $|z| \leq R$, the series

$$\sum_{\nu=m+1}^{\infty} [h_\nu(z) - g_\nu(z)]$$

is (absolutely and) *uniformly* convergent. Since its terms are regular for $|z| \leq R$ (because the poles of the $h_\nu(z)$ with $\nu > m$ lie *outside* the circle $|z| = R$), it defines there a regular function which we shall denote by $F_m(z)$. Then evidently

$$M_0(z) - h_0(z) + \sum_{\nu=1}^{m} [h_\nu(z) - g_\nu(z)] + F_m(z)$$

is also an analytic function which is regular in the circle with radius R about the origin as center, with the exception of those points z_ν in this circle which are poles with principal parts $h_\nu(z)$. The same is valid for every finite region, because R was completely arbitrary; and hence, $M_0(z)$ is a meromorphic function with the required properties.

From the proof it follows that it is sufficient to take the degree n_ν of the polynomial $g_\nu(z)$ (the sum of the first n_ν terms of the power series for $h_\nu(z)$) so large, that, having chosen an arbitrary $R > 0$, the terms $|h_\nu(z) - g_\nu(z)|$ for all $|z| \leq R$ finally (i.e., for all sufficiently large ν) remain less than the terms of a convergent series of positive terms.

Exercises. 1. Does the Mittag-Leffler theorem still hold if the prescribed principal parts contain an infinite number of negative powers? How do the theorem and its proof read then?

2. In connection with Mittag-Leffler's theorem, can one assign the *ascending* part of the Laurent expansion —or at least a finite number of its terms—at all points z_ν (or at several, or one)?

3. The Mittag-Leffler theorem, like the Weierstrass theorem (see §2, Ex. 7), can be carried over to the unit circle. Formulate and prove the theorem indicated.

4. Solve Exercise 3, §3 once again, with the means that have now been developed.

§6. Examples of Mittag-Leffler's Theorem

At times, the "convergence-producing terms" $g_\nu(z)$ are not at all necessary; cf. the analogous case in connection with Weierstrass's theorem. Then, of course, the function to be constructed is especially simple. If, e.g., the points $0, 1, 4, \cdots, \nu^2, \cdots$ are to be poles of order unity with respective principal parts $\dfrac{1}{z - \nu^2}$, then

$$M_0(z) = \frac{1}{z} + \sum_{\nu=1}^{\infty} \frac{1}{z - \nu^2} = \sum_{\nu=0}^{\infty} \frac{1}{z - \nu^2}$$

is a solution. For, let $R > 0$ be chosen arbitrarily, and $m > \sqrt{2R}$. Then the series from $\nu = m + 1$ on is evidently uniformly convergent in $|z| \leq R,$[1] which proves our assertion.

We proceed to construct meromorphic functions corresponding to the examples in §3.

1st Example: $\cot \pi z$. The real lattice points are to be poles of order unity with the residue $+1$, and hence, with the principal parts

$$h_\nu(z) = \frac{1}{z - z_\nu}, \qquad (z_0 = 0,\ z_{2\nu-1} = \nu,\ z_{2\nu} = -\nu).$$

For $\nu = 1, 2, 3, \cdots$,

$$h_\nu(z) = -\frac{1}{z_\nu} - \frac{z}{z_\nu^2} - \frac{z^2}{z_\nu^3} - \cdots ,$$

and it suffices to take all $n_\nu = 0$, and hence,

$$g_\nu(z) = -\frac{1}{z_\nu};$$

because then for all sufficiently large ν (namely, for all $\nu > 4R$) and all $|z| \leq R$,

$$|h_\nu(z) - g_\nu(z)| \leq \frac{R}{|z_\nu|(|z_\nu| - R)} < \frac{2R}{|z_\nu|^2},$$

so that the $|h_\nu(z) - g_\nu(z)|$ finally remain less than the terms of an obviously convergent series of positive terms. Consequently, according to the concluding remark of the preceding paragraph, if $G(z)$ is an arbitrary entire function,

---

[1]Because for $\nu > m$ and $|z| \leq R$ we have

$$|1/(z - \nu^2)| \leq 1/(\nu^2 - R) < 1/(\nu^2 - \tfrac{1}{2}\nu^2) = 2/\nu^2.$$

$$M(z) = G(z) + \frac{1}{z} + \sum_{\nu=1}^{\infty} \left[\frac{1}{z - z_\nu} + \frac{1}{z_\nu} \right]$$

$$= G(z) + \frac{1}{z} + \sum_{\nu=1}^{\infty} \left(\left[\frac{1}{z - \nu} + \frac{1}{\nu} \right] \right.$$

$$\left. + \left[\frac{1}{z + \nu} - \frac{1}{\nu} \right] \right)$$

$$= G(z) + \frac{1}{z} + \sum_{\nu=1}^{\infty} \left[\frac{1}{z - \nu} + \frac{1}{z + \nu} \right]$$

is the most general function of the kind required.

The function cot πz also has poles of order unity at the points $0, \pm 1, \pm 2, \cdots$ (cf. I, p. 126, 3). If n is one of them, the residue at this point is

$$\lim_{z \to n} \frac{(z - n) \cos \pi z}{\sin \pi z} = \lim_{z \to n} \frac{(z - n)[(-1)^n + \cdots]}{(-1)^n \pi (z - n) + \cdots} = \frac{1}{\pi},$$

which can be read off immediately from the indicated series-expansion for a neighborhood of the point $z = n$. Hence, the function π cot πz is contained among the functions $M(z)$ which we just constructed.

We have thus arrived at formula (2) of §3 from an entirely different direction. The still undetermined entire function $G(z)$, which there was called $h_0'(z)$, cannot be ascertained solely from the nature and position of the poles. We should, as before, have to make use of special properties of the function in question. However, in determining the product for sin πz, we have already discovered that we have to set $h_0'(z)$, that is, $G(z)$, equal to zero. Therefore

$$\pi \cot \pi z = \frac{1}{z} + \sum_{\nu=1}^{\infty} \left[\frac{1}{z - \nu} + \frac{1}{z + \nu} \right]$$

is the *partial-fractions decomposition of the cotangent-function.*

2d Example: Weierstrass's $\wp$-function. We shall construct a meromorphic function which has a pole of order *two* with the principal part

$$h_\nu(z) = \frac{1}{(z - z_\nu)^2}$$

at each of the lattice points $z_\nu = k\omega + k'\omega'$, ($\nu = 1, 2, 3, \cdots$), described and enumerated in §3, Example 2. For $\nu = 1, 2, 3, \cdots$,

$$h_\nu(z) = \frac{1}{z_\nu^2} \cdot \frac{1}{(1 - (z/z_\nu))^2} = \frac{1}{z_\nu^2} + 2\frac{z}{z_\nu^3} + 3\frac{z^2}{z_\nu^4} + \cdots,$$

and it is again sufficient to take all $n_\nu = 0$, and hence, to take as $g_\nu(z)$ the first term of this expansion. For then we have

$$h_\nu(z) - g_\nu(z) = \frac{1}{(z - z_\nu)^2} - \frac{1}{z_\nu^2} = \frac{2zz_\nu - z^2}{z_\nu^2(z - z_\nu)^2};$$

and consequently, for all $|z| \leq R$, with arbitrary $R > 0$, and all sufficiently large ν (namely, as soon as $|z_\nu| > 2R$),

$$|h_\nu(z) - g_\nu(z)|$$

$$\leq \frac{R(2|z_\nu| + R)}{|z_\nu|^2(|z_\nu| - R)^2} < \frac{3R|z_\nu|}{|z_\nu|^2(\frac{1}{2}|z_\nu|)^2} = \frac{12R}{|z_\nu|^3}$$

This is (according to §3, p. 29) the general term of a convergent series of positive terms. Hence,

$$M_0(z) = \frac{1}{z^2} + \sum_{\nu=1}^{\infty} \left[\frac{1}{(z - z_\nu)^2} - \frac{1}{z_\nu^2} \right]$$

is a meromorphic function of the type required, and the most general function of this type results immediately therefrom.

In the Weierstrassian theory of elliptic functions, this function $M_0(z)$ is called the **Pe-function** belonging to the pair of periods (ω, ω'), and is denoted by

$$\wp(z) = \wp(z \mid \tfrac{1}{2}\omega, \tfrac{1}{2}\omega').$$

Because of the absolute convergence of the series, the order in which the terms appear does not matter. Hence (cf. p. 30), without further establishing the sequence of the lattice points, we can write

$$\wp(z \mid \tfrac{1}{2}\omega, \tfrac{1}{2}\omega')$$

$$= \frac{1}{z^2} + \sum_{k,k'}{}' \left[\frac{1}{(z - k\omega - k'\omega')^2} - \frac{1}{(k\omega + k'\omega')^2} \right].$$

Here k and k' take on, independently of each other, all positive and negative integral values and zero, without, however, being zero *simultaneously*. This last restriction is indicated by the accent after the summation symbol.

This function $\wp(z)$ bears a close relation to the σ-function of §3, similar to that of cotangent to sine. In fact, according to §2, Theorem 8,

$$\frac{\sigma'}{\sigma}(z) = \frac{1}{z} + \sum_{\nu=1}^{\infty} \left[\frac{1}{z - z_\nu} + \frac{1}{z_\nu} + \frac{z}{z_\nu^2} \right],^{[1]}$$

[1] $\dfrac{\sigma'}{\sigma}(z)$ is an abbreviated notation for $\dfrac{\sigma'(z)}{\sigma(z)}$. This function is frequently called the Weierstrass ζ-function, and denoted by $\zeta(z)$. It has, of course, nothing to do with the Riemann ζ-function

and further. by §2, Theorem 9,

$$-\frac{d}{dz}\left(\frac{\sigma'}{\sigma}(z)\right) = \frac{1}{z^2} + \sum_{\nu=1}^{\infty}\left[\frac{1}{(z-z_\nu)^2} - \frac{1}{z_\nu^2}\right] = \wp(z).$$

Hence,

$$\wp(z) = -\frac{d^2}{dz^2}(\log \sigma(z)) = \frac{(\sigma'(z))^2 - \sigma(z)\,\sigma''(z)}{\sigma^2(z)}.$$

The close connection between Examples 1 and 2 of this paragraph and of §3 suggests the possibility of the existence of a direct relationship between the fundamental theorems themselves—the Weierstrass and the Mittag-Leffler. This is indeed the case: one can derive the first from the second (but not conversely). The method is briefly the following:

Let it be required to construct an entire function $G_0(z)$ with zeros z_ν of respective orders α_ν. First, according to Mittag-Leffler's theorem, construct a meromorphic function $M_0(z)$ having *simple* poles with residues α_ν, and, hence, with principal parts $h_\nu(z) = \dfrac{\alpha_\nu}{z - z_\nu}$, at the points z_ν. One finds, then, almost immediately, that $M_0(z)$ is the logarithmic derivative of an entire function, $G_0(z)$, which can be written in the form of an infinite product, and which satisfies the conditions of Weierstrass's theorem.

---

which is treated in Example 4. At the same time, $\dfrac{\sigma'}{\sigma}(z)$, furnishes us with an example of a meromorphic function which has a simple pole with the residue $+1$ at each of the lattice points of our network of parallelograms.

3d Example: The Gamma-function. The function $\Gamma(z)$, mentioned already in connection with Example 3 of Weierstrass's theorem, proved to be the reciprocal of the entire function $K(z)$ which was constructed there. From this we see immediately that

(1) $\Gamma(z)$ **is a meromorphic function** having a simple pole (for the residue see **(7)** below) only at each of the points $0, -1, -2, \cdots$. Moreover, it is the reciprocal of an entire function, and therefore has no zeros.

We develop several further properties of this important function:

(2) For every $z \neq 0, -1, -2, \cdots$,

$$\Gamma(z + 1) = z\Gamma(z).$$

(Functional equation of the Gamma-function)

Proof:

$$\Gamma(z + 1) = \lim_{n \to \infty} \frac{n!n^{z+1}}{(z + 1)(z + 2) \cdots (z + n + 1)}$$

$$= z \cdot \lim_{n \to \infty} \frac{n!n^z}{z(z + 1) \cdots (z + n)} \cdot \frac{n}{z + n + 1}$$

$$= z\Gamma(z), \qquad\qquad\qquad \text{Q. E. D.}$$

(3) For every integer $\nu \geq 0$,

$$\Gamma(\nu + 1) = \nu!;$$

i.e., $\Gamma(z)$ solves the interpolation problem mentioned on p. 31, footnote.

Proof: For $\nu > 0$, according to **(2)**,

$$\Gamma(\nu + 1) = \nu\Gamma(\nu) = \nu(\nu - 1)\Gamma(\nu - 1) = \cdots = \nu!\Gamma(1);$$

and that $\Gamma(1) = 1$ is seen immediately from the Gaussian definition.

(4) For every z we have[1]

$$\lim_{n \to \infty} \frac{\Gamma(z + n + 1)}{n! n^z} = 1.$$

Proof: The Gaussian definition says that, as $n \to \infty$,

$$\frac{n! n^z}{(z + n)(z + n - 1) \cdots (z + 1) z \cdot \Gamma(z)} \to 1.$$

The denominator here is equal to $\Gamma(z + n + 1)$, as one finds by applying the functional equation $n + 1$ times; and this proves the assertion.

(5) For every non-integral z,

$$\Gamma(z)\Gamma(1 - z) = \frac{\pi}{\sin \pi z}.$$

Proof: According to the Gaussian definition, the reciprocal of the left-hand side is equal to

$$\lim_{n \to \infty} \frac{z(z + 1) \cdots (z + n)}{n! n^z}.$$

$$\cdot \frac{(1 - z)(2 - z) \cdots (n + 1 - z)}{n! n^{1-z}}$$

$$= \lim_{n \to \infty} \frac{n + 1 - z}{n} \cdot z(1 - z^2) \left(1 - \frac{z^2}{2^2} \right) \cdots \left(1 - \frac{z^2}{n^2} \right)$$

$$= z \cdot \prod_{\nu = 1}^{\infty} \left(1 - \frac{z^2}{\nu^2} \right) = \frac{\sin \pi z}{\pi}, \qquad \text{Q. E. D.}$$

[1]The functional equation (2) together with this limit relation (4) are *characteristic* for the Γ-function, i.e., there is no analytic function besides $\Gamma(z)$ which satisfies (2) and (4). We must leave the proof of this proposition to the reader.

(6) $\Gamma(\tfrac{1}{2}) = +\sqrt{\pi}.$[1]

Proof: For $z = 1/2$, **(5)** yields $[\Gamma(1/2)]^2 = \pi$, from which the assertion follows immediately (because from the Gaussian definition $\Gamma(1/2)$ is read off as positive).

(7) The respective residues at the poles $-\nu$, ($\nu = 0, 1, 2, \cdots$), established in **(1)**, are

$$a_{-1}^{(\nu)} = \frac{(-1)^{\nu}}{\nu!}.$$

---

[1]This singular result can also be written as follows:

$$\frac{n!\sqrt{n}}{\tfrac{1}{2}(\tfrac{1}{2}+1)\cdots(\tfrac{1}{2}+n)} \to \sqrt{\pi};$$

or, after a very simple transformation:

$$\frac{1\cdot3\cdot5\cdots(2n-1)}{2\cdot4\cdot6\cdots(2n)} \sim \frac{1}{\sqrt{\pi n}}.$$

The symbol "$\sim$" indicates that the limit of the quotient of both sides as $n \to \infty$ is unity; in other words, that both sides are "asymptotically equivalent", as it is customary to say. The left-hand side is none other than the coefficient of z^n in the binomial series

$$\frac{1}{\sqrt{1-z}} = 1 + \tfrac{1}{2}z + \frac{1\cdot3}{2\cdot4}z^2 + \cdots + (-1)^n\binom{-\tfrac{1}{2}}{n}z^n + \cdots;$$

consequently, **(6)** is synonymous with the statement that

$$(-1)^n\binom{-\tfrac{1}{2}}{n} \sim \frac{1}{\sqrt{\pi n}}.$$

Finally, instead of **(6)** we can write:

$$\frac{\pi}{2} = \frac{2}{1}\cdot\frac{2}{3}\cdot\frac{4}{3}\cdot\frac{4}{5} \cdots \frac{2k}{2k-1}\cdot\frac{2k}{2k+1} \cdots,$$

in which form our result proves to be identical with Wallis's product, which is also obtained at once from the sine-product for $z = 1/2$. (Cf. §3, Ex. 1a).)

Proof: According to its meaning, the residue of a simple pole at $z = -\nu$ is obtained by evaluating

$$\lim_{z \to -\nu} (z + \nu)\Gamma(z).$$

By (2), however,

$$\Gamma(z) = \frac{\Gamma(z + 1)}{z} = \cdots = \frac{\Gamma(z + \nu + 1)}{z(z + 1) \cdots (z + \nu)},$$

so that as $z \to -\nu$,

$$(z + \nu)\Gamma(z) \to \frac{\Gamma(1)}{(-\nu)(-\nu + 1) \cdots (-2)(-1)} = \frac{(-1)^\nu}{\nu!},$$

Q. E. D.

4th Example: The Riemann ζ-function. The Zeta-function, whose most important function-theoretical properties were established by Riemann, plays a fundamental role in the analytic theory of numbers.

In all that follows, t^z, for positive t, is understood to be the (single-valued) entire function $e^{z \log t}$, where $\log t$ has its real value. Then the terms of the series (cf. I, §17, Ex. 2α)

$$\sum_{n=1}^{\infty} \frac{1}{n^z}$$

are *entire* functions. For the absolute values of these terms in the closed half-plane $\Re(z) \geq 1 + \delta$ ($\delta > 0$ arbitrary) we have

$$\left| \frac{1}{n^z} \right| = \frac{1}{n^{\Re(z)}} \leq \frac{1}{n^{1+\delta}}.$$

Therefore, by Weierstrass's M-test, the series is uniformly convergent there. According to I, §19, Theorem 3, since $\delta > 0$ was arbitrary, this means that the series represents a regular function in the half-plane $\Re(z) > 1$. It is this function which is called the *Riemann Zeta-function* and denoted by $\zeta(z)$.

Theorem. $\zeta(z)$ *can be continued across the boundary* $\Re(z) = 1$ *of the half-plane* $\Re(z) > 1$, *and proves to be a meromorphic function having the single pole $z = 1$ with the principal part $1/(z - 1)$; i.e., $z = 1$ is a simple pole with the residue* $+1$.[1]

1st Step: Continuation up to the line $\Re(z) = 0$.

With n^{-z}, the functions $\dfrac{1}{n^{z-1}} - \dfrac{1}{(n + 1)^{z-1}}$ for $n = 1, 2, \cdots$ are álso entire functions. Since each of these has a zero at $z = 1$,

$$\frac{1}{z - 1}\left[\frac{1}{n^{z-1}} - \frac{1}{(n + 1)^{z-1}}\right] = \int_0^1 \frac{dt}{(n + t)^z} = \int_n^{n+1} \frac{dt}{t^z}$$

are also *entire* functions. For $\Re(z) \geq 1 + \delta$, the absolute value of the last integral (by I, §11, 5) is less than or equal to $n^{-1-\delta}$. Hence, for the same reason as before,

$$\sum_{n=1}^{\infty} \int_0^1 \frac{dt}{(n + t)^z} = \int_1^{\infty} \frac{dt}{t^z} = \frac{1}{z - 1}$$

[1]Stated somewhat differently: the difference $\zeta(z) - (z - 1)^{-1}$ is an *entire* (transcendental) *function*.

is convergent for $\Re(z) > 1$. If we subtract this from

$$\sum_{n=1}^{\infty} \frac{1}{(n+1)^z} = \zeta(z) - 1$$

and note that

$$\frac{1}{(n+1)^z} - \int_0^1 \frac{dt}{(n+t)^z} = -z \int_0^1 \frac{t\,dt}{(n+t)^{z+1}}$$

—which can be verified at once by means of an integration by parts on the right-hand side,—we have

(a) $$\zeta(z) = 1 + \frac{1}{z-1} - z \sum_{n=1}^{\infty} \int_0^1 \frac{t\,dt}{(n+t)^{z+1}}.$$

Herewith the continuation in question is accomplished. In order to realize this, one need only show, bearing in mind the form of the first two terms on the right, that the third term, or even only that the new series on the right-hand side, represents a regular function for $\Re(z) > 0$. This follows, however, from considerations quite similar to those encountered before: the terms of the series are again *entire* functions (indeed, they arose from the subtraction of such functions!) and in absolute value are less than or equal to $n^{-1-\delta}$ for $\Re(z) \geq \delta$, from which everything follows as above. Consequently, by means of (a), the asserted character of the point $+1$ is made evident; and, moreover, the region of existence of $\zeta(z)$ is extended to the left by a strip of unit width.

In a similar manner, one can repeatedly extend the region of existence to the left by such a strip, and so finally prove the theorem completely. We shall carry out the next two steps.

2d Step: Continuation up to the line $\Re(z) = -1$. Integrating by parts again, one verifies immediately that

$$-z \int_0^1 \frac{t\, dt}{(n+t)^{z+1}}$$

$$= -\frac{z}{2(n+1)^{z+1}} - \frac{z(z+1)}{2} \int_0^1 \frac{t^2\, dt}{(n+t)^{z+2}},$$

and hence

(b) $$\zeta(z) = 1 + \frac{1}{z-1} - \frac{z}{2}\, [\zeta(z+1) - 1]$$

$$- \frac{z(z+1)}{2} \sum_{n=1}^{\infty} \int_0^1 \frac{t^2\, dt}{(n+t)^{z+2}}.$$

According to the result of the first step, the function inside the brackets of the third term is regular for $\Re(z) > -1$, except at $z = 0$ where there is a simple pole. Because of the factor z before the bracket, the third term itself is a regular function for $\Re(z) > -1$, with no exceptions. This also holds for the last term, because the terms of the new series are entire functions which, in absolute value, are less than or equal to $n^{-1-\delta}$ for $\Re(z) > -1 + \delta$,—from which everything follows once more.

3d Step: Continuation up to the line $\Re(z) = -2$. Another integration by parts gives

$$\int_0^1 \frac{t^2\, dt}{(n+t)^{z+2}} = \frac{1}{3(n+1)^{z+2}} + \frac{z+2}{3} \int_0^1 \frac{t^3\, dt}{(n+t)^{z+3}},$$

and hence

(c) $\zeta(z) = 1 + \dfrac{1}{z-1} - \dfrac{z}{2!}\,[\zeta(z+1) - 1]$

$\qquad\quad - \dfrac{z(z+1)}{3!}\,[\zeta(z+2) - 1]$

$\qquad\quad - \dfrac{z(z+1)(z+2)}{3!}\,\sum_{n=1}^{\infty}\int_{0}^{1}\dfrac{t^3\,dt}{(n+t)^{z+3}}.$

Considerations closely corresponding to those just dealt with show that the only singularity of $\zeta(z)$ in the half-plane $\Re(z) > -2$ is the simple pole at $+1$ with the principal part $\dfrac{1}{z-1}$.

It is now sufficiently clear how one proves that $\zeta(z) - \dfrac{1}{z-1}$ is regular in the half-plane $\Re(z) > -(k+1)$ if it has already been shown that it is regular in the half-plane $\Re(z) > -k$, $(k = 2, 3, \cdots)$. The truth of all our assertions concerning $\zeta(z)$ is thereby established.[1]

[1]That $\zeta(z) - (z-1)^{-1}$ is a *transcendental* entire function follows, e.g., thus: From the series for $\zeta(z)$, we see immediately that $\lim_{x \to +\infty} \zeta(x) = 1$. If the difference in question were a *rational* entire function, it would be *identically* equal to one, so that $\zeta(z) = 1 + (z-1)^{-1}$. That this is false, however, is seen for $z = 2$.

A full treatment of the Riemann ζ-function, including all number-theoretical applications, is to be found in E. Landau, *Handbuch der Lehre von der Verteilung der Primzahlen*, 2 vols., Leipzig, 1909, and E. Landau, *Vorlesungen über Zahlentheorie*, 3 vols., Leipzig, 1927. See also A. E. Ingham, *The Distribution of Prime Numbers*, Cambridge Tracts, No. 30, 1932; E. C. Titchmarsh, *The Zeta-function of Riemann*, Cambridge Tracts, No. 26, 1930.

Exercises. 1. Find the Mittag-Leffler partial-fractions-expansion for each of the following meromorphic functions:

a) $\tan z$;   b) $\dfrac{1}{\sin z}$;   c) $\dfrac{\pi}{\cos (\pi/2)z}$

d) $\dfrac{1}{e^z - 1}$ e) $\dfrac{1}{e^z + 1}$;   f) $\dfrac{1}{\cos z - \sin z}$.

2. The sequence of functions

$$f_n(z) = \frac{n!n^z}{z(z + 1) \cdots (z + n)}$$

is *uniformly* convergent in every bounded, closed region containing none of the points $0, -1, -2, \cdots$.

3. Let z_1 and z_2 be two points distinct from $0, -1, -2, \cdots$. Determine

$$\lim_{n \to \infty} \frac{z_1(z_1 + 1) \cdots (z_1 + n)}{z_2(z_2 + 1) \cdots (z_2 + n)}.$$

4. The entire function $K(z)$ defined in the text has the following representation as an integral:

$$K(z) = \frac{1}{\Gamma(z)} = \frac{1}{2\pi i} \int_k e^t t^{-z} \, dt,$$

where k denotes a path which begins on the left at infinity, proceeds close *below* the negative real axis to a neighborhood of the origin, turns about this point in the positive direction, and then returns to infinity close *above* the negative real axis. (Pay attention to the multiple-valuedness of t^{-z}—as a function of t, for fixed z; details concerning this in sec. II.)

5. Carry out in .detail the derivation, sketched in the text, of Weierstrass's theorem from Mittag-Leffler's.

6. Prove the fact· mentioned on p. 49, footnote, that the Γ-function is characterized uniquely by the two properties **2** and **4** (pp. 48-9).

7. In connection with §2, Ex.4a, show that the Riemann ζ-function has no zeros in the half-plane $\Re(z) > 1$.

CHAPTER 3

PERIODIC FUNCTIONS

§7. The Periods of Analytic Functions

Definition. *An analytic function $f(z)$ is said to be periodic if there exists a non-zero number Ω such that for every z of the domain of regularity of $f(z)$, $z + \Omega$ also belongs to this domain, and*

(1) $$f(z + \Omega) = f(z).$$

Every such number Ω is called a period of $f(z)$.

Among the elementary functions, e^z and the trigonometric functions are periodic. tan z has, e.g., the period -7π; 13π and 2π are also periods of tan z.

In order to confine our attention to what is most important, we shall assume in the following that $f(z)$, except for possible isolated singularities, is single-valued and regular in the entire plane (so that, in particular, entire and meromorphic functions come under consideration). On the other hand, $f(z)$ is not to reduce to a constant, since otherwise equation (1) would be trivial.

If in (1) we replace z by $z + \Omega$, we see that, along with Ω, 2Ω is also a period of the function. The following more general theorem is just as easy to prove:

Theorem 1. *The sum and difference of two periods of a function are also periods of the same;[1] if n, $n' \gtreqless 0$ denote*

---

[1]If $f(z)$ is periodic, the number 0 is also classed with its periods; this is to be noted here and in the following.

any integers, then, along with Ω, *all numbers* $n\Omega$ *are periods, and, along with* Ω *and* Ω', *all numbers* $n\Omega + n'\Omega'$ *are periods.*

We now suppose the points corresponding to all the periods of a function, the so-called "period-points", to be marked in the plane. Then we have the important

Theorem 2. *The set of period-points of a single-valued function has no finite limit point.*

Proof: Otherwise every neighborhood of this limit point would contain an infinite number of period-points, and hence, also such period-points having an arbitrarily small distance (= absolute value of the difference) from each other. Then, according to Theorem 1, there would exist periods with arbitrarily small absolute values, and one could determine a sequence of periods $\Omega_1, \Omega_2, \cdots$ such that $\Omega_n \to 0$. If, now, z_0 is an arbitrary regular point of $f(z)$, and $f(z_0) = a$, then for every $n = 1, 2, 3, \cdots$ we should have

$$f(z_0 + \Omega_n) = f(z_0) = a,$$

implying the existence of a-points in every neighborhood of z_0. But this is impossible, according to I, §21, Theorem 1, since we have assumed that $f(z)$ is not constant. The theorem thus proved can be formulated as follows:

Theorem 3. *A single-valued function cannot have arbitrarily small periods.*

From these theorems, which provide us with a first orientation, we can immediately derive some important consequences.

Let $f(z)$ have the period Ω. The numbers $n\Omega$, which, according to Theorem 1, are also periods, all lie on the line L passing through 0 and Ω, and there constitute a set of equidistant points (see Fig. 2). Suppose there exists a further period-point on L (e.g., tan z with the period -7π has, in addition to the periods $-7\pi n$, the period 3π). It must be of the form

$$n\Omega + \theta\Omega, \qquad (n \text{ integral, } 0 < \theta < 1).$$

Then, by Theorem 1, $\theta\Omega$ itself is a period. Thus, if there exist any periods at all on L besides the periods $n\Omega$, then there are some between 0 and Ω. But there can be only a finite number of these (because of Theorem 2), so that one of them lies nearest to 0. We shall call this one ω,[1] and we now have

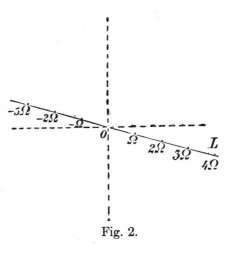

Fig. 2.

Theorem 4. *Every period on L is of the form*

$$n\omega, \qquad (n = 0, \pm 1, \pm 2, \cdots).$$

Proof: By Theorem 1, all the numbers $n\omega$ are periods. If there were one on L besides these, it would be of the form $n\omega + \theta\omega$, $(0 < \theta < 1)$. But then $\theta\omega$, i.e., a point on L between 0 and ω, would also be a period, contrary to the assumption that ω is the period-point on L nearest to 0.

[1]If no further period lies between 0 and Ω, we set $\omega = \Omega$.

By this procedure, ω on L is fully determined except for (the unessential) sign, and is called a **primitive period** of $f(z)$.

The function $\tan z$ has, e.g., the period $\Omega = -7\pi$; L here is the axis of reals. There are six additional periods of $\tan z$ between 0 and -7π, namely, $-\pi$, -2π, $\cdots$, -6π, of which $-\pi$ is the one nearest to 0. Hence, $-\pi$ is a *primitive period* of $\tan z$, so that *all* periods lying on L (i.e. in this case, all *real* periods) have the form $-n\pi$.

If the function has no periods besides the periods $n\omega$ found in this manner, the function is said to be **simply periodic.** In the other case, the periods do not all lie on one straight line, but form, rather, a *plane* point set. We acquire an insight into the nature of this set in the following manner.

Since, according to Theorem 2, there are only a finite number of period-points in any circle, there must exist a *smallest* circle, with center 0, on which there are one or more periods (distinct from 0) (see Fig. 3). We call one of these ω; it is necessarily a primitive period of the function, and on the line L through 0 and ω there are precisely the periods $n\omega$ and no others. We suppose them to be erased for the moment. Then again there exists a smallest circle, with center 0, on which there are one or more of the remaining periods. Let us call that one of them (there are certainly only a finite number) ω' which we first meet in describing this circle in the positive direction if we begin at that half of the line L which extends from 0 to ω. Then the following theorem completely settles the question as to the distribution of the period-points:

Theorem 5. *All period-points of the function are given by*

$$n\omega + n'\omega', \qquad \begin{cases} n = 0, \pm 1, \pm 2, \cdots \\ n' = 0, \pm 1, \pm 2, \cdots \end{cases}$$

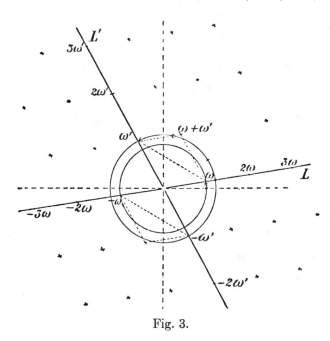

Fig. 3.

Proof: These numbers, according to Theorem 1, are certainly periods. If one existed besides these, it would have to have the form

$$(n + \theta)\omega + (n' + \theta')\omega', \quad 0 \leq \theta \leq 1, 0 \leq \theta' \leq 1,$$

with θ and θ' *not simultaneously* integral (i.e., 0 or 1). But then

$$\Omega = \theta\omega + \theta'\omega'$$

would also be a period. This point would lie in the

parallelogram with the vertices 0, ω, $\omega + \omega'$, ω', without coinciding with any of the vertices. According to the method used to determine ω and ω', it certainly could not lie in the triangle 0, ω, ω', and would therefore have to lie in the other half of the parallelogram in question. But, with Ω,

$$\Omega' = -\Omega + \omega + \omega'$$

would also be a period; and it would be in the triangle 0, ω, ω', where, however, as just shown, there can be no period. Hence, the assumption that there exist periods other than the points $n\omega + n'\omega'$ is inadmissible.

Thus, if a single-valued periodic function is not simply periodic, its periods have the position described by Theorem 5. It will be shown presently (p. 78) that such functions exist. A function of this kind is called **doubly periodic,** and we have

Theorem 6. *The periodicity of a single-valued analytic function can be only simple or double; there is no third.*[1]

The numbers ω and ω' are called a **pair of primitive periods** of the function. Since they are not collinear with 0, their ratio is necessarily non-real:[2]

$$\Re\!\left(\frac{\omega'}{i\omega}\right) \gtrless 0.$$

[1] We call k complex numbers ω_1, ω_2, $\cdots$, ω_k *linearly independent*, if no system of k real, integral numbers n_1, n_2, $\cdots$, n_k, not all zero, exists, such that $n_1\omega_1 + n_2\omega_2 + \cdots + n_k\omega_k = 0$. Then this theorem can be stated as follows: *"A single-valued analytic function cannot possess more than two linearly independent periods."*

[2] The imaginary part of the "period-ratio" ω'/ω is actually *positive* for our determination of the pair of primitive periods, because the positive rotation which carries the direction $(0 \cdots \omega)$ into the direction $(0 \cdots \omega')$ is less than π.

Whereas in the case of a simply periodic function a primitive period, apart from the (quite unessential) sign, is uniquely determined, one can determine a pair of primitive periods of a doubly periodic function in various (infinitely many) ways (cf. Figs. 1 and 3, where the *same* set of points of the form $n\omega + n'\omega'$ is obtained with *different* meanings of ω and ω').

Exercises. 1. A (non-constant) rational function cannot be periodic.

2. A (non-constant) single-valued analytic function cannot have 1 and $\sqrt{2}$ as periods.

3. As a supplement to the consideration of p. 61 for determining ω and ω', show that there can be at most two, four, or six periods on the circle (with center 0) on which ω lies. These are then at the ends of a diameter, the vertices of a rectangle, the vertices of a regular hexagon, respectively.

4. If precisely two periods (hence, ω and $-\omega$) lie on the circle referred to in Exercise 3, the selection of ω and ω' described in the text leads to values for which the *period-ratio* $\tau = \omega'/\omega$ satisfies the conditions:

$$| \tau | \geq 1, \qquad -\tfrac{1}{2} < \Re(\tau) \leq \tfrac{1}{2}.$$

These relations are still valid if four or six periods lie on that circle, provided that a suitable one of these periods is denoted by ω. Proof?

§8. Simply Periodic Functions

One can visualize the periodicity of a simply periodic function in the following manner: Through an arbitrary point c of L, e.g., the origin, draw any line L' which does

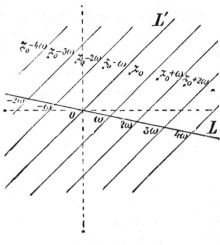

Fig. 4.

not coincide with L (see Fig. 4), and draw parallels to L' through all the points $c + n\omega$. The z-plane is thus divided into strips, which are called **period-strips**. The equation

$$f(z + n\omega) = f(z)$$

now means simply that the function $f(z)$ has the same values at "*congruent*" points, i.e., points that have congruent positions in pairs of strips[1] (cf. Fig. 4, where several points congruent to z_0 have been marked). Consequently, a periodic function exhausts its entire domain of values in one of the strips, even if one regards—as we shall always do in the sequel—only one of the two boundaries as belonging to the strip. In every other strip, all the values, and hence all regular and

---

[1] Two points z_0 and z_1 are called *congruent*, if their difference $z_1 - z_0$ is a period $n\omega$.

singular properties, occur once more: what holds for a point z_0 also holds for every other congruent point.

e^z has the primitive period $2\pi i$. L here is the axis of imaginaries, and for L' we can take the axis of reals. Then the region $0 \leq \Im(z) < 2\pi$, for example, is a period-strip. $\sin z$ has $\omega = 2\pi$ for a primitive period. Hence, if γ denotes any real number, the region $\gamma \leq \Re(z) < \gamma + 2\pi$ can be taken as a period-strip.

It is therefore sufficient to examine a simply periodic function in only one of the strips in order to get to know it completely. At the same time, it is useful to observe that it is no restriction to assume that ω has the particular value $+1$. For, if we set $z = \omega z'$, then $f(z) = f(\omega z')$ goes over into a function which, as a function of z', evidently has the primitive period $+1$.[1] Then precisely the *real integers* are the periods of the function, L is the axis of reals, and for L' we can now take the axis of imaginaries, so that the figure of the plane divided into period-strips becomes especially clear.

The nature of the function $f(z)$ is perceived more distinctly if we make use of the following artifice: We introduce a new variable ζ instead of z by setting

(a) $$z = \frac{1}{2\pi i} \log \zeta, \qquad \zeta = e^{2\pi i z},$$

and consider the function $\varphi(\zeta)$ defined by

$$f(z) = f\left(\frac{1}{2\pi i} \log \zeta\right) = \varphi(\zeta).$$

---

[1] In the foregoing, we have already written very often $e^{2\pi i z}$, $\sin 2\pi z$, $\cot \pi z$, etc. in order to make $+1$ a period of the function in question and thereby give the expansions a simpler form.

Since

$$f(z) = \varphi(\zeta) = \varphi(e^{2\pi i z}),$$

this can also be interpreted thus: The given periodic function will not be regarded or represented as a function of z itself, *but as a function of $e^{2\pi i z}$*.

Since log ζ (cf. I, §26, 1) is an infinitely multiple-valued function, it would seem that $\varphi(\zeta)$ is also a multiple-valued function, and hence, that our investigation is rendered more difficult. However, due to the fact that all values of log ζ result from a particular one by the addition of all integral multiples of $2\pi i$, all values of $\dfrac{1}{2\pi i}$ log ζ differ from one another by integral (real) numbers, and hence, only by periods of $f(z)$. The various determinations of log ζ thus furnish *congruent* values of z, and $\varphi(\zeta)$ therefore proves to be a single-valued function. The multiple-valuedness of the logarithm is just compensated for by the periodicity of the function $f(z)$.

What is the region of existence of $\varphi(\zeta)$, and what is the nature of its singularities? Since log ζ is singular at 0 and ∞ (but at no other point), these two points may be singular for $\varphi(\zeta)$. Other singularities, however, are not introduced by the function log. Apart from these two points, $\varphi(\zeta)$ can be singular in those, and only those, points ζ which, by virtue of (a), correspond to singular points z of $f(z)$. But now a considerable simplification takes place. For, if z_0 is singular for $f(z)$, so are all points of the form $z_0 + k$ with arbitrary integral $k \gtrless 0$. Only *one* singular point

$$\zeta_0 = e^{2\pi i z_0} = e^{2\pi i (z_0 + k)}$$

of the function $\varphi(\zeta)$ corresponds to this entire *set* of singular points of $f(z)$. $\varphi(\zeta)$ thus takes over, so to speak, the singularities of only *one* strip of $f(z)$.

We summarize this result in

Theorem 1. *Every one of the considered simply periodic functions $f(z)$ having the primitive period 1 can be regarded or represented as a single-valued function $\varphi(\zeta)$ of $\zeta = e^{2\pi i z}$. This new function is, in general, of a simpler nature than $f(z)$; for, by virtue of* (a), *only one singularity of $\varphi(\zeta)$ corresponds to every set of congruent singularities of $f(z)$. $\varphi(\zeta)$ is regular except for these and the possibly newly appearing singular points 0 and ∞.*[1]

Examples 1. For the sine-function we have, as is well known,

$$\sin 2\pi z = \frac{1}{2i}\,(e^{2\pi i z} - e^{-2\pi i z}) = \frac{1}{2i}\left(\zeta - \frac{1}{\zeta}\right).$$

$\varphi(\zeta)$ in this case is a very simple rational function.

2. Likewise, the rational function $\varphi(\zeta) = \dfrac{1}{2}\left(\zeta + \dfrac{1}{\zeta}\right)$ corresponds to the function $\cos 2\pi z$.

3. Similarly, we have

$$\tan \pi z = -i\,\frac{\zeta - 1}{\zeta + 1}, \qquad \cot \pi z = +i\,\frac{\zeta + 1}{\zeta - 1}.$$

(As it should, the *single* pole $\zeta = e^0 = +1$, for example, corresponds to the simple poles of the function $\cot \pi z$ at $0, \pm 1, \pm 2, \cdots$.)

With these agreements, it is now easy to derive a

[1]This result holds for every periodic function, and hence, also for the *doubly periodic* functions treated in the next paragraph.

form of expansion which is valid for all our functions $f(z)$. Since $f(z)$ is regular everywhere except for isolated singularities, one can in various ways, by means of parallels to the axis of reals, cut a rectangle out of the period-strip such that $f(z)$ is regular in its interior and on its vertical boundaries. If y_1 and y_2 ($> y_1$) are the ordinates of the aforementioned parallels, $f(z)$ is regular in the entire strip (P) which runs parallel to the axis of reals and is characterized by

$$y_1 < \Im(z) < y_2.$$

Consequently, $\varphi(\zeta)$ is regular provided that z in $\zeta = e^{2\pi i z}$ satisfies the condition just stated, i.e., provided that

$$e^{-2\pi y_2} = r_2 < \mid \zeta \mid < r_1 = e^{-2\pi y_1}. \ [1]$$

Thus, $\varphi(\zeta)$ is single-valued and regular in an annulus determined by the rectangle which was cut out, and can therefore be expanded in one, and only one, Laurent series

$$\varphi(\zeta) = \sum_{n=-\infty}^{+\infty} a_n \zeta^n, \qquad (r_2 < \mid \zeta \mid < r_1),$$

in that ring. This series converges (see I, §30, p. 120) in the broadest ring that can be formed from the hitherto existing one by concentric contraction of the inner circle and expansion of the outer circle, and which is still devoid of singular points. If for ζ we substitute its value in terms of z, we obtain

Theorem 2. *Let $f(z)$ be a single-valued function with the primitive period* 1. *Then, to every strip (P) which is*

[1] $\left| e^{2\pi i z} \right| = \left| e^{2\pi i (x+iy)} \right| = e^{-2\pi y}.$

parallel to the axis of reals and is devoid of singular points, there corresponds one, and only one, expansion of the form

$$f(z) = \sum_{n=-\infty}^{+\infty} a_n e^{2n\pi i z}.$$

This series converges in the broadest strip that can be formed from (P) by a translation of its boundaries upward and downward, and which is still devoid of singular points. Outside this strip the series is divergent.

If this strip contains the axis of reals in its interior— which can always be brought about if the axis has no singular point on it,—we are led from this to the Fourier expansion of real analytic functions by substituting $e^{2\pi i z} = \cos 2\pi z + i \sin 2\pi z$. However, at this point we cannot enter further into this matter.

We must also be content to remark that from every other type of expansion of the single-valued function $\varphi(\zeta)$—ordinary power series, product expansion, partial-fractions expansion—we can, of course, derive a corresponding representation of $f(z)$ as a function of $e^{2\pi i z}$

Of the periodic functions, we have up to now met with only the elementary functions; we shall first get to know others in the next paragraph. For several we have obtained above (p. 68) the corresponding functions $\varphi(\zeta)$, which turned out to be particularly simple, namely: *rational.*[1] We conclude from this, that those simply periodic functions $f(z)$ for which the corresponding function $\varphi(\zeta)$ is rational are especially simple, but also especially important. Moreover, the class of these functions—*the class of rational functions of* $e^{2\pi i z}$—is

[1] The function ζ naturally corresponds to the exponential function itself.

governed by particularly beautiful and typical laws. We shall derive a few of these.

Theorem 3. *Every function $f(z)$ of this class possesses an algebraic addition-theorem; i.e., for variable z_1 and z_2, $f(z_1 + z_2)$ can be expressed algebraically in terms of $f(z_1)$ and $f(z_2)$.*

(For example, for the sine-function, if we make $+1$ the primitive period, we have the addition-theorem written in algebraic form:

$$\sin 2\pi(z_1 + z_2)$$
$$= \sin 2\pi z_1 \cdot \sqrt{1 - \sin^2 2\pi z_2} + \sin 2\pi z_2 \cdot \sqrt{1 - \sin^2 2\pi z_1}).$$

Proof: By hypothesis, $f(z) = R(\zeta)$, where R denotes a *rational* function of its argument. Consequently, if we set $e^{2\pi i z_1} = \zeta_1$, $e^{2\pi i z_2} = \zeta_2$, and hence $e^{2\pi i (z_1 + z_2)} = \zeta_1 \zeta_2$, we have

$$f(z_1) = R(\zeta_1), \quad f(z_2) = R(\zeta_2), \quad f(z_1 + z_2) = R(\zeta_1 \zeta_2).$$

It is possible to eliminate ζ_1 and ζ_2 algebraically from these three rational equations in ζ_1 and ζ_2, so that the algebraic addition-theorem follows immediately.

Theorem 4. *Between any pair of functions $f_1(z)$ and $f_2(z)$ of our class, there exists an algebraic relation.*

Proof: From the hypothesis that

$$f_1(z) = R_1(\zeta) \quad \text{and} \quad f_2(z) = R_2(\zeta),$$

the assertion results immediately from the elimination of ζ.

With $f(z) = \varphi(\zeta)$, the function $f'(z)$, since it is equal to $2\pi i \zeta \varphi'(\zeta)$, also belongs to our class. If we apply the preceding theorem to this function, we obtain

Theorem 5. *Every function* $w = f(z)$ *of our class satisfies an algebraic differential equation of the simple form*[1]

$$w' = \frac{dw}{dz} = A(w),$$

where $A(w)$ *denotes an algebraic function.*

Since it follows from this—under suitable restriction of the variability of w—that

$$dz = \frac{dw}{A(w)}, \qquad z = \int_{w_0}^{w} \frac{dw}{A(w)} = F(w),$$

we can state finally:

Theorem 6. *Every such function* $w = f(z)$ *is the inverse of the integral* $z = F(w)$ *of an algebraic function.*[2]

Exercises. 1. Do there exist (single-valued) simply periodic functions having prescribed zeros (with prescribed orders) in the period-strip? If so, how are they to be set up explicitly?

2. Do there exist (single-valued) simply periodic functions having prescribed isolated singular points in the period-strip? For given principal parts, how can they be set up explicitly?

3. On what region in the w-plane does the function $w = e^{(2\pi i/\omega)z}$ map the fundamental parallelogram of a network of parallelograms in the z-plane determined by (ω, ω')?

---

[1] It is of the first order and does not contain the independent variable.

[2] The reader should verify all of these theorems, and carry out their proofs, with $e^{2\pi iz}$ and the trigonometric functions.

§9. Doubly Periodic Functions; in Particular, Elliptic Functions

The periodicity of a doubly periodic function $f(z)$ can be visualized in a manner analogous to the case of simply periodic functions. If ω and ω' are a pair of primitive periods of $f(z)$, we draw (cf. Fig. 1 or 3) parallels to L through all points $n'\omega'$, and parallels to L' through all points $n\omega$. The entire plane is hereby divided into a network of parallelograms whose lattice points are precisely the period-points $n\omega + n'\omega'$. Let us imagine *this network to be drawn, or any other resulting from an arbitrary translation of this one.* Then the double periodicity of the function evidently means that $f(z)$ assumes the same value, or exhibits the same singularity, at *congruent* points, i.e., now, at points having congruent positions (whose difference is therefore a period) in different parallelograms of the same network. Every one of these parallelograms (that is, every parallelogram with the vertices a, $a + \omega$, $a + \omega + \omega'$, $a + \omega'$; a arbitrary) is called a (indeed, also *the*) period-parallelogram of the function $f(z)$, and we say that $f(z)$ is a doubly periodic function belonging to this period-parallelogram.

In order to get to know such a doubly periodic function completely, it is therefore sufficient to study it "in the period-parallelogram"; e.g., in the so-called "fundamental parallelogram" with the vertices 0, ω, $\omega + \omega'$, ω'. Every regular or singular property which the function possesses at a point z_0 is also found at every one of a set of points $z_0 + n\omega + n'\omega'$, which form the lattice points of one of our parallelogram networks.

The following theorem immediately follows from this remark:

Theorem 1. *There exists no (non-constant) doubly periodic entire function.* **(First Liouville theorem.)**

Proof: As an entire function, $f(z)$ is bounded in every finite region. Consequently, a relation of the form $|f(z)| < K$, where K denotes a suitable constant, is valid for all the points of a period-parallelogram. But then $|f(z)| < K$ for all other points, and hence, in the entire plane. From this it follows, according to I, §28, 1, that $f(z)$ is a constant. We had excluded this trivial case, however.

Thus, a (non-constant) doubly periodic function has at least one singular point in the period-parallelogram. The doubly periodic functions are classified according to the nature of these singular points, which, of course, are the same for every parallelogram.

Definition. *A doubly periodic function which has no singularities other than poles[1] in the period-parallelogram, or in other words: a meromorphic doubly periodic function, is called an* **elliptic function** *belonging to this period-parallelogram.*

We shall further concern ourselves with only these functions in the following. First, we should be entitled to inquire whether such functions exist at all,[2] for we have not encountered any up to now. Since we shall see presently, however, that the function $\wp(z)$, constructed

---

[1] The number of poles in the period-parallelogram then is necessarily finite.

[2] Their discovery, which goes back to Abel and Jacobi, was a very important scientific event.

in §6, Example 2, is an example of a doubly periodic function, we shall not consider the existence question at this moment.

According to the definition, an elliptic function has but a finite number of poles in the period-parallelogram. If we wish to enumerate these, we must make suitable agreements regarding the attachment of the boundaries to the parallelogram. We stipulate that *only the vertex a and the two sides emanating from it, exclusive of their other terminal points, shall be considered as belonging to the period-parallelogram with the vertices a, $a + \omega$, $a + \omega + \omega'$, $a + \omega'$.* With this agreement, it is obvious that for every point of the plane there is always one, and only one, congruent point in an arbitrary one of the period-parallelograms.

It now has a unique meaning to speak of the poles of an elliptic function "in the period-parallelogram." On this we base the

Definition. *The sum of the orders of the poles of an elliptic function in the period-parallelogram is called the* **order** *of the elliptic function.*

The first Liouville theorem can then be stated also as follows:

Theorem 1a. *There exists no (non-constant) elliptic function of order zero.*

Immediately obvious is the following theorem for arbitrary doubly periodic functions:

Theorem 2. *The sum, difference, product, and quotient of two doubly periodic functions $f_1(z)$ and $f_2(z)$ with the pair of primitive periods (ω, ω'), as well as the derivative*

of such a function, are also periodic with the periods ω and ω'. (However, these do not necessarily constitute a pair of primitive periods for the new function; also, this function may be a constant.) If f_1 and f_2 are elliptic, so is the new function.

In connection with Theorem 1, this immediately leads further to

Theorem 3. *If two elliptic functions belonging to the same period-parallelogram have there the same poles with the same respective principal parts, then the functions differ by only an additive constant.*

For, their difference is an elliptic function of order zero.

There is the following important theorem concerning the residues at the poles:

Theorem 4. *The sum of the residues at the poles of an elliptic function in the period-parallelogram*[1] *is equal to zero.* (**Second Liouville theorem.**)

Proof: According to I, §33, the sum in question—apart from the factor $2\pi i$—is given by the integral $\int f(z)\,dz$ taken along the boundary of the parallelogram in the positive sense, provided that no pole lies on the boundary. If this should not be the case for a first choice of the period-parallelogram, it can always be realized at once by means of a sufficiently small translation (e.g., in the direction of a diagonal), which obviously does not have any influence on the present assertion. We may therefore assume that that condition is already

---

[1]As the proof will show, for this theorem to be valid it is not necessary that ω and ω' be a pair of *primitive* periods of $f(z)$.

satisfied to begin with. If, then, a is the vertex associated with this parallelogram,

$$\int_a^{a+\omega} f(z)\, dz = \int_{a+\omega}^{a+\omega+\omega'} f(z)\, dz + \int_{a+\omega+\omega'}^{a+\omega'} f(z)\, dz$$

$$+ \int_{a+\omega+\omega'}^{a+\omega'} f(z)\, dz + \int_{a+\omega'}^{a} f(z)\, dz.$$

Now, we see immediately that the first and third integrals, as well as the second and fourth, differ only in sign; and hence, that their sum is zero. For if we replace z by $z + \omega'$ in the third integral, it becomes

$$\int_{a+\omega}^{a} f(z + \omega')\, dz = -\int_a^{a+\omega} f(z)\, dz,$$

which is minus the first. The proof for the second and fourth integrals is similar. Hence, the sum of the residues is zero, Q. E. D.

From this theorem follows

Theorem 5. *There exists no elliptic function of the first order.*

Proof: It could have only *one* pole of the first order in the period-parallelogram. If its residue were c, the sum of the residues there would also equal c. By the preceding theorem, we should have $c = 0$; i.e., such a pole cannot exist at all.

According to Theorem 2, with $f(z)$, the function $f'(z)/f(z)$ is also an elliptic function with the periods ω and ω'. If we apply Theorem 4 to this function, bearing in mind the proofs of I, §33, Theorems 2 and 3, we obtain

Theorem 6. *The number of zeros of an elliptic function in the period-parallelogram is equal to the number of its poles there—hence, equal to its order* **(Third Liouville theorem.)**

If we apply this theorem to $f(z) - a$, we obtain, finally, the following theorem, which completely settles the question as to the domain of values of an elliptic function:

Theorem 7. *In the period-parallelogram, an elliptic function of order m takes on every value, each precisely m times.*

After these general considerations, we now turn to the actual setting up of several elliptic functions, and the investigation of their most important properties. As we have already stated, the function $\wp(z)$, constructed in §6, Example 2, is an elliptic function. We first prove that this is true of its derivative. Since this derivative can be obtained by differentiating the series for $\wp(z)$ term by term, we have immediately

$$\wp'(z) = -\frac{2}{z^3} - 2 \sum_{\nu=1}^{\infty} \frac{1}{(z - z_\nu)^3};$$

for which we can write, since z_0 was used to denote the origin,

$$\wp'(z) = -2 \sum_{\nu=0}^{\infty} \frac{1}{(z - z_\nu)^3} = -2 \sum_{k,k'} \frac{1}{(z - k\omega - k'\omega')^3},$$

where, in the last series, k and k' take on *all* integral values $\gtreqless 0$ independently of each other and in any order. But if k runs over *all* the integers $\gtreqless 0$, so does $(k - 1)$. Hence, if we substitute $z + \omega$ for z,

$$\wp'(z + \omega) = -2 \sum_{k,k'} \frac{1}{(z - (k - 1)\omega - k'\omega')^3}$$

is actually the same series. Consequently,

$$\wp'(z + \omega) = \wp'(z),$$

and in precisely the same manner we can show that

$$\wp'(z + \omega') = \wp'(z).$$

Herewith is proved the double periodicity of $\wp'(z)$, and hence, the existence of doubly periodic functions in general.

We can show still more precisely, that ω and ω' constitute a pair of *primitive* periods for $\wp'(z)$, and hence, that the numbers $z_\nu - k\omega + k'\omega'$ represent *all* of its periods. For, if Ω is any period of $\wp'(z)$, then $\wp'(z + \Omega) = \wp'(z)$ for every point of the domain of regularity of $\wp'(z)$. If we allow z to approach one of the lattice points, z_ν, (i.e., one of the poles of $\wp'(z)$), $\wp'(z)$ and therefore also $\wp'(z + \Omega)$ become infinitely large. Hence, $z_\nu + \Omega$, too, must be a pole of $\wp'(z)$, and consequently itself a lattice point, z_μ. According to this, $\Omega = z_\mu - z_\nu$, and so Ω is also of the form $k\omega + k'\omega'$, Q. E. D.

Now that we have this result, it is very easy to prove that $\wp(z)$ *itself is a doubly periodic function* with the same pair of primitive periods. By what we just proved, we have

$$\wp'(z + \omega) - \wp'(z) = 0,$$

and hence

$$\wp(z + \omega) - \wp(z) = c,$$

where c denotes a certain constant. We show that this constant must equal zero. The representation of $\wp(z)$ obtained in §6 implies, first of all, that $\wp(-z) = \wp(z)$ For, k and k' take on, independently of each other, all integral values in any order, without being zero simultaneously. We may therefore replace the letters k and k' by $-k$ and $-k'$, respectively. From this remark it follows (because of the exponent 2) that the series is not affected by changing z to $-z$. Hence, $\wp(z)$ is indeed an *even* function. If in $\wp(z + \omega) - \wp(z) = c$ we now substitute the value $z = -\tfrac{1}{2}\omega$, we obtain, as asserted, $\wp(\tfrac{1}{2}\omega) - \wp(-\tfrac{1}{2}\omega) = 0 = c$. Hence,

$$\wp(z + \omega) = \wp(z).$$

Since it is found, in an entirely similar manner, that

$$\wp(z + \omega') = \wp(z),$$

we have established the double periodicity of $\wp(z)$. That ω and ω' constitute a pair of *primitive* periods for this function too follows from the corresponding result for $\wp'(z)$ and from the fact that, in general,

$$\wp(z + \Omega) = \wp(z) \quad \text{implies} \quad \wp'(z + \Omega) - \wp'(z);$$

i.e., from the fact that $\wp'(z)$ can have no other periods than those of $\wp'(z)$. Since, finally, $\wp(z)$ and $\wp'(z)$ are meromorphic functions, we have, to sum up:

Theorem 8. *There exist doubly periodic functions; in particular, elliptic functions; and in fact, such functions* *possessing prescribed pairs of primitive periods. A first example thereof is furnished by Weierstrass's $\wp$-function.*

This $\wp$-function must, for diverse reasons, be regarded as the simplest elliptic function. For, since there exist no elliptic functions of order zero or one, only those of the second order come under consideration as the simplest. That one of these will be regarded as the simplest, which has precisely one pole—call it ζ—of the second order, with the simplest possible principal part, $1/(z - \zeta)^2$, in the period-parallelogram. If, in addition, this pole in the fundamental parallelogram lies at "the" vertex $\zeta = 0$, we are led directly to the function $\wp(z)$,—apart from only an additive constant which comes into question because of Theorem 3.

Even this constant—in itself unimportant—is most easy to determine: About every one of its poles, $\wp(z)$ can be expanded in a Laurent series. For a neighborhood of the origin, this series is of the form

$$\wp(z) = \frac{1}{z^2} + c_0 + c_2 z^2 + c_4 z^4 + \cdots .\ ^1$$

In this expression, the constant c_0 has the value zero, as we immediately infer from

$$\wp(z) - \frac{1}{z^2} = \sum_{\nu=1}^{\infty} \left[\frac{1}{(z - z_\nu)^2} - \frac{1}{z_\nu^2} \right]$$

for $z = 0$.

This "simplest" elliptic function plays a predominant

---

[1] Because the principal part is equal to $1/(z - \zeta)^2 = 1/z^2$, and, since $\wp(z)$ has already been shown to be an *even* function, no odd powers can appear.

role in the Weierstrassian theory of elliptic functions,[1] analogous to that of the exponential function in the theory of simply periodic functions. We can only illustrate this importance of the function $\wp(z)$ by means of several samples of the theory. We first prove (cf. §8, Theorem 5) the fundamental

Theorem 9. *The function $w = \wp(z)$ satisfies the algebraic differential equation of the first order*:

$$\left(\frac{dw}{dz}\right)^2 = (w')^2 = 4w^3 - g_2 w - g_3;$$

where g_2 and g_3 denote certain constants, the so-called "invariants of the $\wp$-function", which are determined by ω and ω' alone. The independent variable does not appear in the differential equation, and w' is an algebraic function of w.

Proof: From the series for $\wp(z) - z^{-2}$ which we just used, and setting[2]

$$\sum_{\nu=1}^{\infty} \frac{1}{z_\nu^n} = s_n$$

$(n \geq 3)$ for brevity, it follows, due to Weierstrass's double-series theorem (I, p. 83) and the fact that

$$\frac{1}{(z - z_\nu)^2} = \frac{1}{z_\nu^2} + 2\frac{z}{z_\nu^3} + 3\frac{z^2}{z_\nu^4} + \cdots,$$

---

[1] In the (older) Jacobian theory, a function plays the role of the simplest, which, in the period-parallelogram, has two separate poles with residues (then necessarily) differing only in sign. These poles lie at the midpoints of the sides belonging to the fundamental parallelogram.

[2] According to p. 29, the series are absolutely convergent for $n \geq 3$.

that

$$\wp(z) = \frac{1}{z^2} + 2s_3 z + 3s_4 z^2 + 4s_5 z^3 + 5s_6 z^4 + \cdots .$$

If, now, we follow Weierstrass and set

$$60s_4 = g_2 = 60 \sum_{k,k'}{}' \frac{1}{(k\omega + k'\omega')^4},$$

$$140s_6 = g_3 = 140 \sum_{k,k'}{}' \frac{1}{(k\omega + k'\omega')^6},$$

noting that the s_n with odd subscripts must equal zero,[1] then the initial terms of the desired expansion are

$$\wp(z) = \frac{1}{z^2} + \frac{g_2}{20} z^2 + \frac{g_3}{28} z^4 + \cdots$$

From this it follows, further, that

$$\wp'(z) = -\frac{2}{z^3} + \frac{g_2}{10} z + \frac{g_3}{7} z^3 + \cdots ,$$

$$(\wp(z))^3 = \frac{1}{z^6} + \frac{3}{20} g_2 \cdot \frac{1}{z^2} + \frac{3}{28} g_3 + \cdots$$

With these expansions, form the function

$$(\wp'(z))^2 - (4\wp^3(z) - g_2 \, \wp(z) - g_3).$$

According to Theorem 2, this function is, first of all, an elliptic function with the same periods and no other poles. It can be verified immediately, that no negative

[1]Since the lattice points of the network, taken in suitable pairs, differ only in sign, the corresponding pairs of terms of the series for s_n, with n odd, are annulled.

powers appear in its expansion for a neighborhood of the origin. Therefore, by Theorem 1a, the function reduces to a constant,—and in fact, to zero; since calculation shows further, that the constant term is also missing in the expansion. Hence, as asserted,

$$\wp'^2 = 4\wp^3 - g_2\wp - g_3,$$

where g_2, g_3 have the above-mentioned values.[1] It is well-known that, conversely, the function which satisfies such a differential equation is also uniquely determined as soon as two corresponding values of the

[1]This result has implications in many directions. We call attention at this point to the following:

By virtue of the equation

$$y^2 - (4x^3 - g_2x - g_3) = 0,$$

y is defined as a multiple-valued function of x, and x is defined as a multiple-valued function of y. (Here x and y, and afterwards t, denote complex numbers.) Each is an algebraic function of the other (further details of this are given in ch. 5); or, carrying over some notions from the "real" domain, we say that the equation defines an *algebraic curve*. Our foregoing result now shows that this "curve" has a parametric representation in

$$x = \wp(t),\ y = \wp'(t).$$

This fact, namely, that we have found two *single-valued* functions of a parameter t which yield precisely the same curve as the given implicit equation by means of which each of the variables x and y is defined as a *multiple-valued* function of the other, is expressed by saying that we have **uniformized** the curve. Thus, here we have uniformized a particular algebraic curve of the third degree with the aid of the $\wp$-function. We are dealing with a simpler case of uniformization when we represent the "circle" $x^2 + y^2 - 1 = 0$, with the aid of the trigonometric functions, in the form $x = \cos t$, $y = \sin t$. The problem of uniformization here indicated plays an important role in the modern theory of functions.

variables are known. For the $\wp$-function, $z = 0$ and $w = \infty$, e.g., constitute such a pair of values. Consequently, we infer from

$$\frac{dw}{dz} = \sqrt{4w^3 - g_2w - g_3},$$

that

$$z = \int\limits_{\infty}^{w} \frac{dw}{\sqrt{4w^3 - g_2w - g_3}};$$

and hence, that $w = \wp(z)$ is the *inverse function*, or the *inverse*, of the function $z = z(w)$ defined by this integral.[1] Such an integral—more generally, every integral whose integrand is a rational function of the variable itself and of the square root of a polynomial in that variable of the third or fourth degree whose roots are all distinct—is called an **elliptic integral** (for the rather superficial reason that it first appeared in connection with the rectification of an arc of an ellipse). It was in the inverses of such functions defined by means of elliptic integrals, that Abel and Jacobi first discovered doubly periodic functions, which they therefore gave the name "elliptic functions".

[1]We started from the periods ω and ω', constructed the $\wp$-function belonging to these, and found the above-mentioned differential equation for this function, in which the invariants g_2 and g_3 were determined by ω and ω' alone. Now, the so-called *problem of inversion* is to determine whether, conversely, if g_2 and g_3 are assigned arbitrarily, the inverse of the function $z = z(w)$, given as the integral above, is a $\wp$-function whose invariants are the numbers g_2 and g_3.

We now set as our last aim, to prove a theorem which gives us, in a certain sense, a survey of the totality of elliptic functions:

Theorem 10. *Every elliptic function with the periods ω and ω' can be represented rationally in terms of the function $\wp(z \mid \frac{1}{2}\omega, \frac{1}{2}\omega')$ and its derivatives, provided that ω/ω' is not real.*[1]

The proof gives us at the same time an opportunity to become acquainted with several further important properties of our functions.

From $\wp(z + \omega) = \wp(z + \omega') = \wp(z)$ it follows (cf. p. 47) that

$$(1) \quad \frac{\sigma'}{\sigma}(z + \omega) = \frac{\sigma'}{\sigma}(z) + \eta;\ \frac{\sigma'}{\sigma}(z + \omega') = \frac{\sigma'}{\sigma}(z) + \eta',$$

if η and η' denote suitable constants.[2] And from this we find, further, that

$$\sigma(z + \omega) = c \cdot e^{\eta z}\sigma(z),\ \sigma(z + \omega') = c' \cdot e^{\eta' z}\sigma(z).$$

Since $\sigma(z)$ is also an odd function, the new constants are again obtained by letting $z = -\frac{1}{2}\omega$, and we have more precisely:

[1]That, conversely, every function which is a rational combination of $\wp(z)$ and its derivatives is an elliptic function with the periods ω and ω', is self-evident according to Theorem 2.

[2]Since $\dfrac{\sigma'}{\sigma}(z)$ is an odd function, we find, on setting $z = -\frac{1}{2}\omega$,

that $\frac{1}{2}\eta = \dfrac{\sigma'}{\sigma}(\frac{1}{2}\omega)$. Hence, according to p. 46, η can be calculated by means of a series, and η', of course, likewise.

$$(2) \quad \begin{cases} \sigma(z + \omega) = -e^{\eta(z+\frac{1}{2}\omega)}\sigma(z), \\ \sigma(z + \omega') = -e^{\eta'(z+\frac{1}{2}\omega')}\sigma(z). \end{cases}$$

By means of the same two integration steps, we get the initial terms of the Laurent expansions for a neighborhood of the origin from those for $\wp(z)$:

$$(3) \qquad \frac{\sigma'}{\sigma}(z) = \frac{1}{z} + b_3 z^3 + \cdots,$$

$$(4) \qquad \sigma(z) = z + d_5 z^5 + \cdots,$$

where we need not know the easily calculated coefficients b_3, d_5, and all the following ones.

Now we see on the basis of (2), that the function

$$-\frac{\sigma(z - a)\, \sigma(z + a)}{\sigma^2(z) \cdot \sigma^2(a)},$$

in which a is an arbitrary point distinct from the lattice points, admits the periods ω and ω'; and hence, since it has only the single pole $z = 0$ in the fundamental parallelogram, that it represents an elliptic function, $\varphi(z)$, belonging to this parallelogram. Since, now,

$$\sigma(z \pm a) = \pm\, \sigma(a) + \sigma'(a) \cdot z \pm \tfrac{1}{2}\sigma''(a) \cdot z^2 + \cdots,$$

the beginning of the expansion of $\varphi(z)$ for a neighborhood of the origin reads

$$\varphi(z) = \frac{1}{z^2} + \frac{\sigma(a) \cdot \sigma''(a) - (\sigma'(a))^2}{\sigma^2(a)} + c_2 z^2 + \cdots$$

$$= \frac{1}{z^2} - \wp(a) + e_2 z^2 + \cdots \qquad \text{(see p. 47)},$$

where again we need not know the coefficients $e_2, \cdots$.

These first few terms show, however, that $\varphi(z) - \wp(z) + \wp(a)$ is the constant 0; for, it is an elliptic function of order zero, which vanishes for $z = 0$. If, in order to emphasize the freedom in the choice of a, we write z' instead of a, we have the fundamental formula

$$\wp(z) - \wp(z') = -\frac{\sigma(z - z')\,\sigma(z + z')}{\sigma^2(z)\cdot\sigma^2(z')}.$$

If we differentiate this formula logarithmically, first with respect to z, then with respect to z', and add, we find the so-called *addition-theorem for the function* $\dfrac{\sigma'}{\sigma}\,(z)$

$$\frac{\sigma'}{\sigma}\,(z + z') = \frac{\sigma'}{\sigma}\,(z) + \frac{\sigma'}{\sigma}\,(z') + \frac{1}{2}\,\frac{\wp'(z) - \wp'(z')}{\wp(z) - \wp(z')}.$$

From this we obtain the addition-theorem for the function $\wp(z)$ by differentiating once more with respect to z:

$$\wp(z + z') = \wp(z) - \frac{1}{2}\frac{\partial}{\partial z}\left(\frac{\wp'(z) - \wp'(z')}{\wp(z) - \wp(z')}\right).^{[1]}$$

With these preliminaries, we are now in a position to prove Theorem 10 as follows:

[1]If we perform the differentiation and make use of the differential equation for $\wp(z)$—from which we get $\wp''(z) = 6\wp^2(z) - (g_2/2)$,—we can also write the addition-theorem in the form:

$$\wp(z + z') = \frac{[2\wp(z)\wp(z') - (g_2/2)][\wp(z) + \wp(z')] - g_3 - \wp'(z)\wp'(z')}{2[\wp(z) - \wp(z')]^2},$$

in which it is seen to be an *algebraic addition-theorem* (cf. §8, Theorem 3), because $\wp'(z)$ and $\wp'(z')$ are expressible algebraically in terms of $\wp(z)$ and $\wp(z')$.

Let the given elliptic function $f(z)$ have the k poles $\zeta_1, \zeta_2, \cdots, \zeta_k$ with the respective principal parts

$$h_\nu(z) = \frac{a_{-1}^{(\nu)}}{z - \zeta_\nu} + \cdots + \frac{a_{-\alpha_\nu}^{(\nu)}}{(z - \zeta_\nu)^{\alpha_\nu}}$$

in the period-parallelogram with the vertices 0, ω, $\omega + \omega'$, ω'. Now, since each of the functions

$$\frac{\sigma'}{\sigma}(z), \quad \wp(z), \quad -\frac{1}{2!}\wp'(z), \quad +\frac{1}{3!}\wp''(z), \cdots$$

has at the origin a pole with the simple principal part

$$\frac{1}{z}, \quad \frac{1}{z^2}, \quad \frac{1}{z^3}, \quad \frac{1}{z^4}, \cdots,$$

respectively, we see at a glance, that the function

$$H_\nu(z) = a_{-1}^{(\nu)} \cdot \frac{\sigma'}{\sigma}(z - z_\nu) + a_{-2}^{(\nu)} \cdot \wp(z - \zeta_\nu) -$$

$$- \frac{a_{-3}^{(\nu)}}{2!}\wp'(z - \zeta_\nu) + \cdots + \frac{(-1)^{\alpha_\nu} a_{-\alpha_\nu}^{(\nu)}}{(\alpha_\nu - 1)!}\wp^{(\alpha_\nu - 2)}(z - \zeta_\nu)$$

is a meromorphic function which has a pole with the principal part $h_\nu(z)$ at ζ_ν.

According to the addition-theorem for the $\wp$-function, $\wp(z - \zeta_\nu)$, and hence also its derivatives, can be expressed rationally in terms of $\wp(z)$ and its derivatives. Consequently, bearing in mind the formulation of our theorem, and replacing $\wp(z)$, $\wp'(z)$, $\cdots$ for brevity by $\wp$, $\wp'$, $\cdots$, we can write more simply

$$H_\nu(z) = a_{-1}^{(\nu)} \cdot \frac{\sigma'}{\sigma}(z - \zeta_\nu) + R_\nu(\wp, \wp', \cdots),$$

where by R_ν we mean a rational function of its arguments. If we now add together the H_ν, ($\nu = 1, 2, \cdots$, k), the sum

$$a_{-1}^{(1)} \frac{\sigma'}{\sigma} (z - \zeta_1) + a_{-1}^{(2)} \cdot \frac{\sigma'}{\sigma} (z - \zeta_2) + \cdots$$

$$+ a_{-1}^{(k)} \cdot \frac{\sigma'}{\sigma} (z - \zeta_k)$$

appears. By virtue of the addition-theorem for $\frac{\sigma'}{\sigma} (z)$, this sum can be replaced by

$$(a_{-1}^{(1)} + a_{-1}^{(2)} + \cdots + a_{-1}^{(k)}) \frac{\sigma'}{\sigma} (z)$$

$$- \left[a_{-1}^{(1)} \frac{\sigma'}{\sigma} (\zeta_1) + \cdots + a_{-1}^{(k)} \frac{\sigma'}{\sigma} (\zeta_k) \right]$$

$$+ \{\text{a rational function of } \wp(z) \text{ and } \wp'(z)\}.\text{[1]}$$

By Theorem 4, the parenthesis is equal to zero; and since the bracket is constant, the sum in question is a rational function of $\wp$ and $\wp'$. We set it equal to $R_0(\wp, \wp')$, and now we have

$$H_1(z) + \cdots + H_k(z) = R_0(\wp, \wp') + \sum_{\nu=1}^{k} R_\nu(\wp, \wp', \cdots).$$

The sum of all the $H_\nu(z)$ is thus a rational function of $\wp(z)$ and its derivatives, and hence, in particular, an elliptic function (despite the individually *non*elliptic

[1] If one $\zeta_\nu = 0$, the corresponding term in the bracket is missing.

terms $\frac{\sigma'}{\sigma}(z - \zeta_\nu)!$). If we subtract it from $f(z)$, this function obviously loses all its poles, and therefore reduces to a constant, C_0, so that we obtain the representation

$$f(z) = C_0 + R_0(\wp, \wp') + \sum_{\nu=1}^{k} R_\nu(\wp, \wp', \cdots)$$

$$= R(\wp, \wp', \cdots),$$

which is our theorem.[1]

We shall have to be satisfied, within the limits of our little book, with these samples taken from the very extensive theory of elliptic functions.

Exercises. 1. If ω is positive and ω' is a positive pure imaginary, then the $\wp$-function $\wp(z \mid \tfrac{1}{2}\omega, \tfrac{1}{2}\omega')$ is *real* on the boundary of the fundamental parallelogram (which in this case is a rectangle). Proof?

2. Under the conditions of the preceding exercise, on what region of the w-plane does $w = \wp(z)$ map the fundamental rectangle?

3. In connection with the preceding exercise, effect the conformal representation of a given rectangle on the unit circle.

---

[1]From $\wp'^2 = 4\wp^3 - g_2\wp - g_3$ we get, by differentiating:
$\wp'' = 6\wp^2 - (g_2/2)$, and further:
$\wp''' = 12\wp\wp'$,
$\wp'''' = 12\wp'^2 + 12\wp\wp'' = 120\wp^3 - 18g_2\wp - 12g_3$,

etc. We see, in general, that *all* higher derivatives of $\wp(z)$ can be expressed as *polynomials* in $\wp(z)$ and $\wp'(z)$. If we make use of this result, we can sharpen Theorem 10 to the effect that **all** *elliptic functions can be expressed rationally in terms of* $\wp(z)$ *and* $\wp'(z)$.

4. Is $e^{\wp(z)}$ an elliptic function?

5. Carry out in detail the proof of the algebraic addition-theorem for the $\wp$-function as indicated in the footnote on p. 88.

6. Show that, in the fundamental parallelogram, $\wp'(z)$ has the simple zeros $\frac{1}{2}\omega$, $\frac{1}{2}(\omega + \omega')$, $\frac{1}{2}\omega'$,—and no others.

MULTIPLE-VALUED FUNCTIONS

CHAPTER 4

ROOT AND LOGARITHM

§10. Prefatory Remarks Concerning Multiple-valued Functions and Riemann Surfaces

We return now to the developments in I, ch. 8, particularly those of §24. There we saw how one can, in general, derive more and more new functional elements from a first such element, given, say, in the form of a power series, by means of analytic continuations—of which every single one is always absolutely unique and necessary,—and thereby enlarge the domain of existence of the function. We imagined this to proceed as far as possible. The complete analytic function resulting in this manner from one element was defined to be *single-valued*, when its behavior at every single particular point, z_0, is always the same, independent of the path along which one may reach it by analytic continuations; in other words, when every point, z_0, which has *once* belonged to the interior of a circle of convergence can never constitute an obstacle for *any* continuation, and when it is always made to correspond to the same functional value in the process of any continuation. Then all points of the z-plane are separated unambiguously into the regular and the nonregular, and to every regular point there is made to correspond one, and only one, functional value. The totality of regular points forms a region in the sense of I, §4, the *region of existence* of the function, whose points are the "bearers" of the functional values of the single-valued function $w = f(z)$.[1]

These functions are naturally easier to handle[2] than others, and we have therefore dealt with them up to now almost exclusively.

All this is altered when we have a *multiple-valued* function before us, in which case the above-mentioned condition is not fulfilled in the continuation process. Then it is possible for several distinct functional values to correspond, as a result of different continuations, to one and the same point, and one and the same point may actually prove to be regular in one continuation and singular in another. Provisionally, we imagine that with every z are associated *all* those functional values

[1]Imagine the functional values w to be written on, pinned to, or in some other way affixed to the proper point z of the region of existence.

It is often advantageous to add the poles to the region of existence and to make them bearers of the value ∞.

[2]Nevertheless, it need not (cf. the remark in I, pp. 93-94) be possible to obtain a single-valued function completely with the aid of a single expression, as was indeed the case with the entire and the meromorphic functions.

The region of existence may also have the most varied and complicated forms; for there is actually the following **Theorem:** *For every region* ⑤ (see I, p. 18), *there are analytic functions having precisely the region* ⑤ *for their region of existence, and hence are not continuable beyond* ⑤ (see, e.g., L. Bieberbach, *Lehrbuch der Funktionentheorie*, vol. I, New York, 1945, pp. 295-296).

which it acquires in the course of all possible continuations; in addition, if occasion arises, the designation "singular" is applied to it if it proves to be singular in *any* continuation.

Let $w = F(z)$ be the functional configuration obtained in this way. Then the symbol $w = F(z)$, for a given z, no longer has a uniquely determined sense, but rather can have several (a finite or an infinite number of) meanings. Examples are $\sqrt{z}$ and $\log z$, which we have already treated somewhat more closely.

The *main question*, then, is, in general, the following—at first formulated quite loosely: *How does one keep the various determinations of a multiple-valued function apart, how does one bring order and insight into its domain of values?*

In definite individual cases this question will usually assume the following form: If a problem which by its nature must have a fully unique solution is solved with the aid of a multiple-valued function, *which of its determinations is the one to use?*

A few examples, in which we make use of the already somewhat familiar function log, will throw light on this formulation of the question and its difficulty.

1. Let (cf. p. 3, footnote) $H(z) = a_0 + a_1 z + \cdots$ be an entire function with no zeros; let b_0 be the principal value of the logarithm of $a_0 (\neq 0)$. Then, as we have seen, in virtue of the condition that $h(0)$ shall equal b_0, $h(z) = \log H(z)$ becomes a well-defined entire function. For every z, $h(z)$ is *one* logarithm of $H(z)$; e.g., $h(1)$ is

one value of $\log \left(\sum_{n=0}^{\infty} a_n \right)$. *Which one is it?*

2. Let the non-zero complex numbers a and b be distinct, and let k be a path connecting them but not passing through the origin. Then $\int_{\substack{k \\ a}}^{b} \dfrac{dz}{z}$ is a definite complex number. Since $\dfrac{d}{dz} \log z = \dfrac{1}{z}$, it is contained among the infinitely many values of $\log b - \log a$. *Which one of these values is it?*

3. Let the single-valued function $f(z)$ be regular on, and in the interior of, the simple closed path C (cf. I, §33, Theorem 2), and let $f(z) \neq 0$ *along C.* Let a denote any point of C. Then, since $\dfrac{d}{dz} (\log f(z)) = \dfrac{f'(z)}{f(z)}$,

$$\frac{1}{2\pi i} \int_C \frac{f'(z)}{f(z)}\, dz = \frac{\log f(a) - \log f(a)}{2\pi i}$$

is certainly an integer $\gtreqless 0$. *What is its value?*

4. Finally, we wish to show that a point may be regular for some continuations, singular for others:

$\sqrt{-2\pi i}$ has two values; we select a definite one of these and denote it by c_0. Then we see immediately, that there exists one, and only one, power series, $\sum c_n(z - 1)^n$, which begins with c_0, converges in a neighborhood of $+1$, and for which

$$[c_0 + c_1(z - 1) + \cdots]^2$$

$$= -2\pi i + (z - 1) - \frac{(z - 1)^2}{2} + \cdots ;$$

which, consequently, in a few words, represents a value of $\sqrt{\text{Log } z - 2\pi i}$, where Log z denotes the principal value of log z. $z = +1$ is, of course, a *regular* point for this functional element. If we now imagine this element to be continued along, say, the unit circle in the positive sense—which is obviously possible,—then, on returning, the point $+1$ turns out to be *singular*; for, Log z has been increased by $2\pi i$, and $\sqrt{\text{Log } z}$ is evidently not regular at $z = +1$ any more.[1]

We can now formulate the question somewhat more sharply in the following manner: Let $w = f_0(z)$ be a regular element, at $z = z_0$, of the multiple-valued function $w = F(z)$, and let it be continuable along the path k extending from z_0 to ζ. Then we land at ζ with a definite one of the functional values $F(\zeta)$. *Which one of these values is it?*

For the present we shall give merely a cursory presentation of the method for overcoming these difficulties. Only after we have examined several examples more closely in this and the next chapter shall we seek a general answer in the last chapter.

The functional element from which we proceed may be assumed to be a power series. We imagine its circle of convergence to be cut out of paper, and its points to be made bearers of the (unique) functional values of the element. If, now, we continue the initial element by

---

[1]Something similar actually occurs already in the domain of real numbers. $x^3 + y^3 - 3axy = 0$ represents the so-called folium of Descartes. y is triple-valued for all $0 < x < a \sqrt[3]{4}$. The upper two of the three arcs are singular at $x = a \sqrt[3]{4}$ (the differential quotient is ∞) and "are joined there"; the other arc remains regular.

means of a second power series, we also think of its circle of convergence as being cut out and pasted in the proper position on the first disk. (We hereby obtain a figure like that in I, p. 100, Fig. 7b.) The parts pasted together are bearers of the same functional values, and are accordingly counted henceforth as a single sheet covered *once* with values. If we succeed in carrying out another continuation, we paste the new disk on in an entirely similar manner, etc. Each new disk is pasted on the preceding one, from which it was obtained by means of (*eo ipso* single-valued) continuation, in the manner described.

Suppose that, after repeated continuation, we arrive with one of the new circles over old territory—i.e., over a circular disk not immediately preceding (cf. I, p. 103 and Fig. 8). Then, the new disk shall be pasted together with the old one when, and only when, both are bearers of the same functional values, or, only so far as both bear the same functional values.[1] If, however, they bear different functional values, let them overlap and remain disconnected. Then two sheets, which are bearers of different—but on each sheet fully unique—functional values, are superimposed on this part of the plane.

We always obey this rule in the future, and we imagine our procedure to be continued as long as possible. Then there results a surface-like configuration which covers the plane with several, in general actually

---

[1] If the function is single-valued, then the united circles gradually fill out the entire region of existence of the function precisely once; and the resulting sheet, together with its affixed functional values, represents the function completely.

infinitely many, *"sheets,"* which can have the most varied forms, and can be joined together in the most varied manners.[1] It is called the **Riemann surface** of the multiple-valued function $w = F(z)$ defined by the initial element. The entire domain of values of $F(z)$ is spread out on it in a completely single-valued manner, to the extent that, on every sheet, every point is the bearer of one, and only one, value. (All possible free boundaries or boundary points of a sheet of this surface are singular for the continuations giving rise to the sheet in question; for details, see below.)

Only after we have illustrated these very general ideas by several transparent examples shall we be able to fully appreciate the advantage of this method of representation.

Note, finally, that it is of course immaterial whether we continue by means of circular disks or by means of any other regions—say in the manner described in I, p. 92—provided only that we adhere to the agreements we have made.

Exercises. 1. Is it possible for a multiple-valued function to have the same value at two superposed points of its Riemann surface? Can it have the same value at *all* points of a neighborhood of two such points?

2. What kind of function (single-valued or multiple-valued) is defined by each of the following formulas:

---

[1]In the course of pasting sheets together, it is sometimes necessary to join two sheets which are separated by others lying between them. We must imagine this to take place without cutting the intermediate sheets, without touching them, so to speak, in the process. This, of course, is impossible for concrete execution, but causes no difficulty for the purely mental construction which we are solely concerned with here.

a) $\sqrt{e^z}$ b) $\sqrt{\cos z}$, $\cos \sqrt{z}$;   c) $\sqrt{1 - \sin^2 z}$;

d) $\sqrt{\wp(z)}$;   e) $\sqrt{\wp(z) - \wp(\tfrac{1}{2}\omega)}$;   f) $\log (e^z)$?

§11. The Riemann Surfaces for $\sqrt[v]{z}$ and $\log z$

1. $w = \sqrt{z}$ can be regarded as the simplest multiple-valued function. We saw in I, §26, 2, that it is possible to continue the real function $\sqrt{x}$, defined and positive for $x > 0$, into the complex domain. For a neighborhood of $+1$, e.g., the continuation is effected by the binomial series

$$[1 + (z - 1)]^{\frac{1}{2}}$$

$$= 1 + \frac{1}{2} (z - 1) - \frac{1}{2 \cdot 4} (z - 1)^2$$

$$+ \frac{1 \cdot 3}{2 \cdot 4 \cdot 6} (z - 1)^3 - + \cdots$$

Proceeding from this element, we can carry out the continuation absolutely unhindered and in a fully unique manner so long as we avoid the negative axis of reals. With reference to the procedure, described in the preceding paragraph, for constructing the Riemann surface, this has the following significance: By joining domains, we can first of all fill out the entire plane which is cut only along the negative axis of reals, and make every one of its points the bearer of a single value of $\sqrt{z}$ which is *uniquely* determined by the initial element that was chosen. We say that, from the entire domain of values of $w = \sqrt{z}$ (where every $z \neq 0$ bears *two* values), we have extracted a *branch* which is regular in the region just mentioned.

Now, since the origin is the only finite singularity (concerning the point ∞, see p. 105), a further, analogous continuation across the edges of this domain is possible. However, if, e.g., we continue downward across the upper bank of the cut, we are no longer permitted to paste together. For, the parts of the region which project from the upper bank into the lower half-plane are now bearers of *different* functional values. These values, as we know, are the values already affixed there multiplied by the factor $e^{2\pi i/2} = -1$; i.e., the first values with opposite sign. The plane is thus covered a second time by these regional segments;[1] and since the origin is the only obstacle to continuation, this second cover-

Fig. 5.

[1]We imagine this second sheet to lie *above* the first.

ing will eventually spread over the whole plane (excluding 0) until it returns to the negative axis of reals, receiving in the process the same values as before with sign changed.

Then the entire plane is overlaid with two sheets which wind around the origin, like a helicoid, in two turns, and on which the entire domain of values of $\sqrt{z}$ is spread out precisely once (cf. Fig. 5). The surface has two free boundaries, one above and one below, extending along the negative axis of reals, across which we can continue the function still further. Assume this to take place once more across the upper bank into the lower half-plane. Then, the attached regional parts, which at first cover the plane a third time, are bearers of values which again are equal to those below them multiplied by the factor (-1); they are thus bearers of the *same* functional values as those in the lower half-plane of the lowermost sheet. We therefore do *not* have a third sheet, but rather, in a few words, we must join the two free edges *by penetrating the intermediate sheet*. But then everything is accomplished at a blow. For now there is no free boundary and no possibility for an analytic continuation any more: we have obtained the *Riemann surface for the function* $w = \sqrt{z}$, on which the entire domain of values of this function is unfolded in a fully unique manner. $\sqrt{z}$ is a *single-valued* function of position on this surface. If we proceed from point to point along any path on the surface, we move in just as single-valued a domain of values as in the case of a single-valued function.[1]

[1]Since, after the first sheet was finished, we could have con-
.. ued the function upward instead of downward, or in both direc-

A (non-zero) point of the surface now is no longer determined by z alone; it is also necessary to specify the sheet on which it lies. Since numbering the two sheets is, naturally, of no consequence, it is usually arranged so that the surface is thought of as being built up in some arbitrary, but henceforth fixed, manner. Then a point of the same is determined uniquely by naming z and the value of $\sqrt{z}$ affixed there. Finally, the point 0, round which the two sheets hang together, is also added to the surface, but is only counted *once*. We let it bear the value 0, and call it a *simple branch-point* of the surface (or a *branch-point of order one*). It is easy to verify that every complex number w is affixed once, and only once, to our Riemann surface.

In a neighborhood of every non-zero point of the Riemann surface, the attached domain of values constitutes a single-valued regular analytic function. Moreover, this neighborhood may be expanded so far as it remains single-sheeted and, consequently, does not contain the origin. If $\mathfrak{G}$ is such a region (e.g., the circle, in one sheet, with center z_0 and radius $\mid z_0 \mid$; or the right half-plane of one sheet; or the plane cut along the

---

tions at the same time, we see that the form and position of the line of penetration of the two sheets is quite unimportant. The penetration is, so to speak, *not there at all*, or, in any case, is only involved in the imperfection of our empirical space-perception. Note, merely, that the winding-surface just has the characteristics that two sheets lie one above the other at *every* point distinct from zero, and that it is capable of returning us to any one of its points (different from zero), taken as starting-point, after a double circuit of the origin. The penetration of the intermediate sheet, which is necessary for our material spatial perception, need not be thought of at all; disregard it completely.

negative axis of reals), we say that it, together with its covering of values, represents a *branch* of the function.

(If, e.g., we denote by C the unit circle described in the positive sense, and if we begin at $+1$ with the meaning of $\sqrt{z}$ developed on p. 100, then accordingly

$$\int_{\substack{C \\ +1}}^{+1} \frac{dz}{\sqrt{z}} = [2\sqrt{z}\,]_{+1}^{+1\,(C)} = -2 -2 = -4.)$$

2. Matters are analogous for the function $w = \sqrt[p]{z}$, $(p > 2)$. The results of I, §26, 2 show, on the basis of similar considerations, that in this case the continuation process, as it was envisaged with the aid of Fig. 5, is not yet complete after two encirclements of the origin. If we continue across the free edges of the surface obtained thus far (say, once more, across the upper bank into the lower half-plane), we see that a third covering of the plane is necessary in order to accommodate the domain of values; etc. Only after p coverings is it apparent[1] that another continuation across the upper bank into the lower half-plane does not lead to a new covering of the latter, but, rather, to the one which is already attached to the lowermost sheet in the lower half-plane. Therefore we penetrate the intermediate $(p - 1)$ sheets and fuse the upper bank of the pth sheet with the lower bank of the first sheet. Therewith every boundary disappears, and the continuation process is now complete.

---

[1]The values associated with the points of the second sheet differ from those of the first by the factor $e^{2\pi i/p}$, with those of the third, by the factor $e^{2(2\pi i/p)}$, $\cdots$, with those of the pth, by the factor $e^{(p-1)(2\pi i/p)}$, and, at the next step, differ by the factor $e^{p(2\pi i/p)} = 1$, i.e., *not at all.*

On this *p-sheeted Riemann surface for the function* $w = \sqrt[p]{z}$, its entire domain of values is uniquely unfolded, and, in addition, all the remarks we just made in the special case $p = 2$ are valid here. The origin is called a *branch-point of order* $(p - 1)$. We now add it to the surface, counting it *once*, and let it bear the value 0.

We indicate briefly, that it is possible to carry out exactly the same considerations using the *sphere* of complex numbers (cf. I, p. 4) instead of the z-plane. We arrive at an analogous two-, p-sheeted covering of the sphere with a *Riemann spherical surface*. The reader will be able to assure himself of this without any difficulty. The branch-point 0 of order $(p - 1)$ is now at the south pole, and we find—and herein lies the advantage of using the *sphere*—that the north pole, i.e., the point ∞, is a point of an entirely analogous nature, namely, a branch-point of order $(p - 1)$. We have before us a p-sheeted Riemann sphere with *two* branch-points; neither is favored above the other on the sphere. Corresponding to the agreements made in the case of 0, we also add the point ∞ to the surface, count it only *once*, however, and let it bear the value ∞.

It can be verified immediately, that *every* complex number w (including 0 and ∞) is affixed to the surface once, and only once.

One's insight into such a fact can be made more vivid in the following way: In studying a function $w = f(z)$, instead of letting the points z *bear* the values w, as we have done up to now, place near the z-plane (or z-sphere) a w-plane (w-sphere), and mark the *point* $w = f(z)$ on it. We call this point, briefly, the *image* of z in virtue of the mapping function $f(z)$. If we displace z

continuously on its sphere (naturally within the domain of existence of $f(z)$), w will also move continuously on its own sphere. Consequently, the mapping itself is said to be continuous. To every point, line, figure on the z-plane or z-sphere there corresponds, on the basis of this continuous representation, an "image" on the w-surface. The nature of this *mapping* must be regarded as *characteristic for the function.*[1]

If we make use of this notion, we can also state our last results as follows: *By means of $w = \sqrt[p]{z}$, the corresponding p-sheeted Riemann sphere is mapped one-to-one on the simple*[2] *(i.e., one-sheeted) w-sphere. To every point of the one configuration corresponds one, and only one, point of the other.*[3]

3. The construction of the Riemann surface for $w = \log z$ is analogous to that for $w = \sqrt[p]{z}$, since here, too, only 0 and ∞ are singular. We begin, say, with the principal value, regular at $+1$: the functional element

[1] We cannot discuss here in greater detail the *properties* of this mapping—its continuity was of course evident,—important as they are for the whole development of the theory of functions. Their investigation forms the content of the little volume of L. Bieberbach, *Einführung in die konforme Abbildung*, 3d ed., Sammlung Göschen No. 768, Berlin and Leipzig, 1937. See also C. Carathéodory, *Conformal Representation*, Cambridge Tracts No. 28, 1932.

[2] The German word is *schlicht*. The term "smooth" is sometimes used in this connection.

[3] If one imagines every point, z, of the z-plane to be connected with its image, w, of the w-plane by means of an invisible thread, then these, in their totality, constitute that "inner bond" of which we spoke in I, p. 103.

$$(z - 1) - \frac{(z - 1)^2}{2} + \frac{(z - 1)^3}{3} - + \cdots .$$

It can (cf. I, §26, 1) be continued uniquely and unhindered over the z-plane cut along the negative axis of reals. We thus obtain a single-valued and regular function in this region; it is the *principal value* or *principal branch* of $\log z$.

We can further continue across the edges of the cut. If this is done, as in the case of $\sqrt[p]{z}$, we get a second sheet (to be thought of as lying *above* the first), whose points bear values equal to the subjacent ones increased by $2\pi i$. The same occurs when we come to a third sheet, etc. The construction of the surface proceeds in exactly the same manner as for $\sqrt[p]{z}$; except that here we never come to an end, because *every* sheet bears values equal to those immediately below increased by $2\pi i$. We therefore suppose that above *every* sheet there lies another one; whereby, in imagination, the continuation process in the upward direction is completed, i.e., a further extension in this direction is impossible.

Now, the lower bank of the first sheet is still free, and we can continue across it (into the upper half-plane). We think of the resulting new sheet as lying *below* the first; it bears values equal to the superjacent (principal) values diminished by $2\pi i$,—new values in any case. This ever-possible continuation process also produces new sheets with new coverings endlessly in the downward direction. For, every sheet bears values equal to those immediately above diminished by $2\pi i$. If we accordingly suppose that below *every* sheet there lies another, we finally, in imagination, complete the process of constructing the surface, which now is not open to extension in any direction.

The entire domain of values of $w = \log z$ is unfolded on this infinite-sheeted Riemann surface in a fully *unique* manner; log z is a *single-valued function of position* on the surface.

If we carry out the same procedure on the sphere instead of the plane—picture an infinite number of spherical shells wound around one another,—we see immediately that the point ∞ (the north pole) is exactly the same sort of point as the point 0. We have before us an *infinite-sheeted Riemann sphere with two branch-points of infinite order*. Such branch-points are *never* added to the surface, and are *never* made to bear functional values.

It remains for us to say a few words about the distribution of the domain of values of log z on our surface. On the first sheet are attached the principal values of $\log z = \int\limits_{1}^{z} \dfrac{d\zeta}{\zeta}$, i.e., those for which the path of integration runs wholly—but arbitrarily—in the plane cut along the negative axis of reals. If, in order to get from $+1$ to z, we proceed, say, first along the positive axis of reals to the point $|z|$, and thence along the circle with center 0 and radius $|z|$ along the shortest path to z, we have

$$\log z = \int\limits_{1}^{|z|} \frac{d\zeta}{\zeta} + \int\limits_{|z|}^{z} \frac{d\zeta}{\zeta} = \int\limits_{1}^{|z|} \frac{dx}{x} + i \int\limits_{0}^{am\,z} d\varphi$$

$$= \text{Log}\,|z| + i\,am\,z,$$

where Log $|z|$ denotes the real logarithm of the positive number $|z|$, and we take

$$- \pi < \text{am } z \leq + \pi.^{1}$$

(The multiple-valuedness of log z accordingly appears as an immediate consequence of the ambiguity of the amplitude of a complex number.) The principal value of $w = \log z$ thus satisfies the condition

$$- \pi < \Im (w) \leq + \pi.$$

The *point* w therefore lies in the strip of the w-plane characterized by precisely this inequality; its width is 2π, and it lies symmetric with respect to the axis of reals. We read off from $z = e^{w}$, that *every* point w of this strip is the image of a $z \neq 0$. Consequently, by means of the principal value of log z, the cut plane, 0 excluded, is mapped one-to-one on the indicated strip of the w-plane. The remaining values of log z, which are to be found on the other sheets, differ from the principal value by only a term of the form $2k\pi i$ with integral $k \gtrless 0$. The corresponding *points* w consequently lie in the strips characterized by

$$(2k - 1) \pi < \Im (z) \leq (2k + 1) \pi,$$

$$(k = \pm 1, \pm 2, \cdots).$$

These are joined to the first strip in an unbroken sequence, and fill out the entire w-plane precisely once.[2]

[1] We add the upper bank (am $z = +\pi$) of the cut to the cut plane in order that every $z \neq 0$ shall lie in it precisely once.

[2] Evidently they are simply the *period-strips* of the function e^{w}. Indeed, according to this, the multiple-valuedness of $w = \log z$, which consists solely in an arbitrary term of the form $2k\pi i$, is precisely the "inverse" phenomenon to the simple periodicity of the inverse function $z = e^{w}$, which has the primitive period $2\pi i$.

We may therefore say that *by means of w = log z, the infinite-sheeted Riemann z-plane with the two branch-points 0 and ∞ (not belonging to it) is mapped one-to-one (and continuously) on the simple w-plane*; or, that every complex number ($\neq \infty$) is attached to one, and only one, point on the Riemann surface for log z.

The questions raised as examples on pp. 95-96 are now easily answered:

1. log $H(z)$ is uniquely defined as follows: The choice of log $H(0)$ means that we begin at the point $H(0)$ of an arbitrarily chosen, but now fixed, sheet of the log-surface. Let us proceed from 0 along two paths, k_1 and k_2, to a point z_0. Then the value of the function H varies from $H(0)$ to $H(z_0)$ along two paths which, since $H(z) \neq 0$, can neither pass through, nor enclose, the point 0. Both, therefore, lead us on the log-surface to one and the same perfectly well-determined point of a perfectly well-determined sheet. There the functional value log $H(z_0)$ stands uniquely affixed.

2. $\int_{\substack{k \\ a}}^{b} \dfrac{dz}{z} = \log b - \log a$ is now to be understood

as follows: Beginning at the point a on any sheet (i.e., choosing log a arbitrarily), we describe the path k *on the log-surface*. Since k must not pass through 0, it leads us to the point b of a perfectly well-determined sheet, and here the value of log b, which alone comes into question, is uniquely attached.

3. Here, too, everything is determined uniquely if we follow the path described on the log-surface by $f(z)$ as z traverses the path C.

As a further application, we prove the following

Theorem. *Let the two functions $f(z)$ and $\varphi(z)$ be regular in the simply connected region $\mathfrak{G}$. Let the simple closed path C lie within $\mathfrak{G}$, and let*

$$f(z) \neq 0, \qquad |f(z)| > |\varphi(z)|$$

along C. Then the two functions $f(z)$ and $f(z) + \varphi(z)$ have the same number of zeros in the subregion of $\mathfrak{G}$ enclosed by C. (**Rouché's Theorem.**)

Proof: According to I, §33, Theorem 2, it is sufficient to show that

$$\int_C \left[\frac{f' + \varphi'}{f + \varphi} - \frac{f'}{f} \right] dz = \int_C \frac{(1 + (\varphi/f))'}{(1 + (\varphi/f))} \, dz = 0.$$

The last integral is equal to $[\log (1 + (\varphi/f))]^{(C)}$, i.e., the difference between the initial value and the terminal value of $\log (1 + (\varphi/f))$ when C is traversed in the positive senso. But, when z describes the path C, the value of $1 + (\varphi/f)$ remains in the right half-plane (actually, inside the circle with center $+1$ and radius 1), because $|\varphi/f| < 1$. Since the (more precisely: *every*) logarithm is fully unique there, the above-mentioned difference must equal zero, Q. E. D.

Exercises. 1. Evaluate the integrals

$$\int_1^i \frac{dz}{\sqrt[p]{z}} \quad \text{and} \quad \int_1^i \log z \, dz,$$

where k denotes the first quadrant of the unit circle that lies on the first-covered sheet of the proper Riemann surface.

2. What values can the integral $\int_1^{z_0} \dfrac{dz}{\sqrt[p]{z}}$ have for an arbitrary path extending from a definite one of the (p distinct) points $+1$ to a definite one of the (p distinct) points $z_0 \neq 0$?

What values can the integral $\int_1^{z_0} \log z \, dz$ have for a corresponding interpretation?

3. Prove the fundamental theorem of algebra, with the aid of Rouché's theorem, by setting

$$\varphi(z) = a_0 + a_1 z + \cdots + a_{n-1} z^{n-1} \text{ and } f(z) = a_n z^n,$$

$$(a_n \neq 0),$$

and choosing for C a circle with a sufficiently large radius.

§12. The Riemann Surfaces for the Functions
$$w = \sqrt{(z - a_1)(z - a_2) \cdots (z - a_k)}$$

The situation as regards the function $w = \sqrt{z - a}$ (a arbitrary) is quite analogous to the case $w = \sqrt{z}$; only, instead of the origin, the point a is the simple branch-point. For the sphere, the difference appears even more unessential, since now, as before, we obtain a two-sheeted Riemann sphere with two branch-points. The branch-point which previously lay at the south pole now lies at a—in all other respects the considerations of the preceding paragraph remain unaltered.

It is but a short step from this surface to the surface

for the function $w = \sqrt{(z - a_1)(z - a_2)}$, where a_1 and a_2 denote arbitrary, but distinct, complex numbers. We again obtain a two-sheeted Riemann sphere having simple branch-points at a_1 and a_2. The point ∞ is now like any other point distinct from a_1 and a_2; i.e., the two sheets pass by each other smoothly at ∞.

We shall derive all this once more, directly, according to the general considerations of §10:

$$w = \sqrt{(z - a_1)(z - a_2)}$$

is regular at all points distinct from a_1 and a_2. If z_0 is such a point, let $\sqrt{z_0 - a_1}$ and $\sqrt{z_0 - a_2}$ be arbitrary, but henceforth fixed, values of these double-valued square roots. Then, with the aid of the binomial series employed on p. 100, we obtain from the regular elements

$$f_\nu(z) = \sqrt{z_0 - a_\nu} \left(1 + \frac{z - z_0}{z_0 - a_\nu}\right)^{\frac{1}{2}}, \qquad (\nu = 1, 2),$$

at z_0 an expansion of w:

$$f = f_1 \cdot f_2 = c_0 + c_1(z - z_0) + c_2(z - z_0)^2 + \cdots,$$

which converges for a neighbourhood of z_0. f constitutes an element of our function. Starting with this element, we shall construct the Riemann surface. We can continue unhindered over the entire plane so long as we avoid the points a_1 and a_2. Since $w = \sqrt{z - a_1} \cdot \sqrt{z - a_2}$, a multiple covering of a point z could only take place if it were reached along two paths beginning at z_0 and surrounding one of the two points a_ν. Let us cut the plane from a_1 to a_2 to ∞, and never pass over this cut during the continuation.[1] Then, by means of

---

[1] The cut can be drawn arbitrarily; but it must not intersect itself, and must avoid the point z_0.

our continuation process, one, and only one, functional value is associated with every point of this cut plane; the latter becomes a bearer of a branch of our function.[1] Now we can further continue across the banks of the cut, and the only question is, whether in so doing we obtain new coverings or not. If we had only the element $f_1(z)$ of the function $\sqrt{z - a_1}$ before us, we should only have to make one cut from a_1 to ∞ (corresponding to the cut along the negative axis of reals in §11), and, in crossing it, $\sqrt{z - a_1}$ would be multiplied by (-1). The same holds for the element $f_2(z)$ of the function $\sqrt{z - a_2}$ if we make a cut from a_2 to ∞. Accordingly, the cut from a_1 to a_2 to ∞ which we just drew can assume this role for both square roots. Let us continue across that part of the cut which extends from a_1 to a_2 (and which is oriented thus). Then, since f_1 goes over into $-f_1$ and f_2 remains unaltered, the regions already covered are overlaid with *new* values, namely, the old ones with sign changed. There results a second sheet— which we imagine to lie, say, *above* the first—with a second covering (differing from the first only in sign). It hangs together *crosswise*, to put it briefly, with the first sheet along the cut-segment in question, a crossing of this segment *always* leading us from one sheet to the other. The second sheet also must be thought of as being cut from a_2 to ∞.

Then *both* sheets at present are cut along $a_2 \cdots \infty$. If we cross this cut in the process of continuation still

[1]We are making use here of the following obvious **Theorem:** *If $f(z) = f_1(z) \cdot f_2(z)$ (identically) for the functional elements f, f_1, f_2, and if it is possible to continue f_1 and f_2, then the product of their continuations is a continuation of $f(z)$ itself.*

possible, *both* of the above-mentioned square roots are multiplied by (-1), so that w remains unaltered; i.e., such a crossing does *not* lead to a new covering, but we must rather join each sheet to itself along this part of the original cut. Therewith all boundaries disappear, and hence no further continuation is possible. Thus, we have obtained once more the above-described two-sheeted Riemann surface with two branch-points.

$\sqrt{(z - a_1)(z - a_2)}$ is a *single-valued* function of position on this surface. We again add the points a_1 and a_2 (each counted only *once*) to the surface and let them bear the value 0, and we likewise add the points ∞ of the two sheets (which are simple there) and let them bear the value ∞. Then *every* complex number w (including ∞) is to be found at *precisely two points* of our Riemann surface.

It is not at all difficult to apply these considerations to the functions

$$w - \sqrt{(z - a_1)(z - a_2) \cdots (z - a_k)},$$

where $k > 2$, and $a_1, a_2, \cdots, a_k$ are distinct, but otherwise arbitrary, complex numbers. One has only to draw a cut, not intersecting itself, from a_1 to a_2 to a_3 to $\cdots$ to ∞,[1] and carry out exactly the same considerations with the k factors $\sqrt{z - a_\nu}$, $(\nu = 1, 2, \cdots, k)$, into which our function can be factored, as we just did for $k = 1, 2$. We find, then, that the two sheets must be joined *crosswise* between a_1 and a_2, and likewise between a_3 and a_4, etc., whereas each sheet must be rejoined *to itself* between a_2 and a_3, a_4 and a_5, $\cdots$. We

[1]If it is advantageous, the numbering of the a_ν may be changed for this purpose.

thus obtain in *both* cases, whether k be odd or even
($2r - 1$ or $2r$, say), a two-sheeted Riemann surface
with the even number, $2r$, of simple branch-points lying
at the points a_ν and—for odd k—also at ∞.[1]

We call attention to the particularly important cases
$k = 3$ and $k = 4$, each of which leads to a two-sheeted
surface with *four* branch-points.

Although these surfaces may seem to be quite analogous to those with two branch-points that were obtained for $k = 1$ and $k = 2$, a fundamental difference
between the two deserves to be stressed. The two-
sheeted sphere with *two* branch-points can be mapped
one-to-one and continuously on the *simple* sphere. It
follows from this,—also from direct consideration,—
that every simple closed curve *on the surface*[2] divides it
into two pieces which are completely separated by this
curve—as is also the case for the simple sphere. It is
customary to express this by assigning the two surfaces
the same *genus*; in the present instance, the *genus zero*.

The two-sheeted sphere with *four* branch-points
(which is obtained for $k = 3$ and $k = 4$) is essentially
different in this respect. Let us draw the curve C (see
Fig. 6), enclosing a_1 and a_2 (but only these two branch-
points), entirely in the upper sheet. Then we can still
connect the two points z_0 and z_1 on the upper sheet,
where z_0 and z_1 lie on opposite sides of C, by means of a
path lying on the surface but not intersecting C. In

[1] For even k, on the other hand, ∞ is an ordinary point, i.e.,
both sheets are simple there, so that ∞ appears on the surface
twice—as was described in detail for $k = 2$.

[2] That is, the curve extends from a point z_0 of some sheet back
to the same point on the same sheet.

Fig. 6 we attempt to make this clear by representing the path in the lower sheet by an interrupted line, in the upper sheet by a continuous line; the cuts along which the sheets have been joined crosswise are dotted.[1] We therefore assign a different genus to this surface, namely, the *genus one*. In general, the surfaces with $2r$

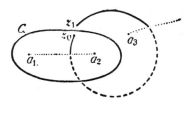

Fig. 6.

branch-points, corresponding to $k = 2r - 1$ and $k = 2r$, are given the *genus* $(r - 1)$.

We can only remark very briefly, that these essentially different connectivities of, e.g., the surfaces for $\sqrt{1 - w^2}$ and $\sqrt{4w^3 - g_2 w - g_3}^2$ constitute the basic reason that the integrals .

$$z = \int_0^w \frac{dw}{\sqrt{1 - w^2}} \text{ and } z = \int_\infty^w \frac{dw}{\sqrt{4w^3 - g_2 w - g_3}}$$

[1]The cut beginning at a_3 leads to ∞ or a_4, according as k is equal to 3 or 4.

[2]Here we must interchange the letters z and w in order to remain in accord with §§8 and 9.

The above roots are, except for a constant factor, obviously of the type dealt with, with $k = 2$ and $k = 3$.

or, more clearly, their inverses

$$w = \sin z \quad \text{and} \quad w = \wp(z)$$

are of such fundamentally different types, namely, simply-, doubly-periodic, respectively.

Exercises. 1. Prove the theorem formulated in the footnote on p. 114.

2. What values can the integral $\displaystyle\int_0^{z_0} \frac{dz}{\sqrt{1 - z^2}}$ have if

the path extends from a definite one of the two points 0, in an arbitrary manner, but avoiding the points ± 1, to a definite one of the points $z_0(\neq \pm 1)$?

Do these integrals still have any meaning for $z_0 = \pm 1$ and $z_0 = \infty$? If so, what do they mean?

CHAPTER 5

ALGEBRAIC FUNCTIONS

§13. Statement of the Problem

All the examples of multiple-valued functions handled thus far, with the exception of log z, have been *algebraic functions*. They are named thus because they result from the solution of an algebraic equation, i.e., an equation of the form $G(z, w) = 0$, where G denotes an entire rational function of z and w. If we imagine G to be arranged in ascending powers of w, it can be written in the form

$$g_0(z) + g_1(z) \cdot w + g_2(z) \cdot w^2 + \cdots + g_m(z) \cdot w^m = 0,$$

where the coefficients $g_\nu(z)$ represent polynomials in z alone.

Now, if a functional element $w = f(z)$ is such, that when substituted for w in such an equation it satisfies the equation *identically*, i.e., that $G(z, f(z)) = 0$ for *all* z of a certain region, we say that $f(z)$ is an element of an *algebraic function* defined by $G(z, w) = 0$. If $m \leq 4$, one can solve the equation, and thereby obtain the function in question explicitly, and examine it. For $m > 4$ this is no longer possible, as is well known. There arises, then, just the problem of whether a function is *at all* defined in a similar manner by such an equation, whether only *one*[1] function is yielded, what the nature of this function is.

First we state the hypotheses precisely. We may assume $G(z, w) = 0$ to be *irreducible*, i.e., not expressible as the product of two polynomials of the same type as G.[2] For obviously the treatment of an equation of the form

$$G_1(z, w) \cdot G_2(z, w) = 0$$

can be replaced by the separate consideration of the equations $G_1 = 0$ and $G_2 = 0$.

If, now, we imagine a particular value z_0 to be substituted for z, we have before us an equation in w, with numerical coefficients, which, in general, will have m distinct roots $w_0^{(1)}$, $w_0^{(2)}$, $\cdots$, $w_0^{(m)}$. An exception takes place only if

a) $g_m(z_0) = 0$, because then the degree of the equation is lowered, or if

b) $G(z_0, w) = 0$ has multiple roots.

This last case can occur if, and only if, a certain expression, the so-called *discriminant* of the equation, which is an entire rational function of the coefficients, vanishes. Furthermore, if $G(z, w)$ is assumed to be irreducible, the discriminant, which we shall denote by $D(z)$, does not vanish identically, but, on the contrary, is a polynomial of a definite degree. (We must assume

---

[1]That more than one function may be defined by such an equation is already demonstrated by so simple an example as $w^2 - z^2 = 0$, which obviously yields *two* functions.

[2]This concept is an absolute one in the case of *two* variables, whereas in the case of *one* variable it has a definite meaning only after the nature of the numerical coefficients has been established.

that these algebraic facts are familiar to the reader.) Consequently, the exceptions mentioned under a) and b) can, in any case, occur *only for a finite number of special values of z*, which we denote by $a_1, a_2, \cdots, a_r$. We shall exclude these "critical points" from consideration for the present. Then we can say that the equation $G(z_0, w) = 0$ has precisely m distinct roots, $w_0^{(1)}, \cdots, w_0^{(m)}$, for every $z = z_0$ distinct from the critical points; z_0 is made the bearer of these roots. Our goal is the following

Theorem. *The m-fold domain of values, which is made to correspond to the points of the plane ("punctured" by the exclusion of the a_r) in virtue of the equation $G(z, w) = 0$, is that of a single m-valued analytic function, $w = F(z)$. Or, more briefly: the equation $G(z, w) = 0$ defines precisely one m-valued regular function, $w = F(z)$, in the punctured plane.*

Functions that can be defined in such a manner are called **algebraic functions.**

We shall prove this theorem, which forms the basis of the theory of algebraic functions, in the next two paragraphs; and beyond that, we shall consider the behavior of $F(z)$ at the critical points, with which we also class the point ∞.

§14. The Analytic Character of the Roots in the Small

Let z_0 be a point which, for the present, is subject solely to the condition that $g_m(z_0) \neq 0$. Then $G(z_0, w) = 0$ has, in any case, m roots, some of which may be multiple roots, however. Let w_0 be such an α-fold root, $1 \leq \alpha \leq m$. Then there is the following

Theorem on the continuity of the roots. *If a circle K_ϵ with a sufficiently small radius $\epsilon > 0$ is described about w_0 as center, then it is possible to draw such a small circle K_δ with radius $\delta = \delta(\epsilon) > 0$ about z_0 as center, that, for every $z_1 \neq z_0$ in K_δ, the equation $G(z_1, w) = 0$ has precisely α distinct roots in K_ϵ.*[1]

Proof: If we set $w = (w - w_0) + w_0$ and arrange in ascending powers of $(w - w_0)$, we may write

$$G(z, w) = \bar{g}_0(z) + \bar{g}_1(z) \cdot (w - w_0) + \cdots$$
$$+ \bar{g}_m(z) \cdot (w - w_0)^m,$$

and for the new coefficients we have

$$\bar{g}_0(z_0) = \bar{g}_1(z_0) = \cdots = \bar{g}_{\alpha-1}(z_0) = 0, \ \bar{g}_\alpha(z_0) \neq 0.$$

Consequently, it is possible, first of all, to describe such a small circle $K_{\delta'}$ with radius $\delta' > 0$ about the point z_0 as center, that $D(z)$ and $\bar{g}_\alpha(z)$ differ from zero within and on the boundary of $K_{\delta'}$ (except, possibly, $D(z)$ at the center z_0, in case $\alpha > 1$). We then set

$$G(z, w) = \bar{g}_\alpha(z) \cdot (w - w_0)^\alpha \cdot [1 + A + B],$$

where

$$A = A(z, w)$$
$$= \frac{\bar{g}_{\alpha+1}}{\bar{g}_\alpha} \cdot (w - w_0) + \cdots + \frac{\bar{g}_m}{\bar{g}_\alpha} \cdot (w - w_0)^{m-\alpha},$$

[1] This theorem, at the same time, gives a deeper interpretation of the multiplicity of a root; for it says that an α-fold root of an equation branches off into α simple roots if the coefficients of the equation are varied a little.

$$B = B(z, w)$$

$$= \frac{\bar{g}_{\alpha-1}}{\bar{g}_\alpha} \cdot \frac{1}{w - w_0} + \cdots + \frac{\bar{g}_0}{\bar{g}_\alpha} \cdot \frac{1}{(w - w_0)^\alpha}.^1$$

Now, let $c > 0$ be the greatest lower bound of $|\bar{g}_\alpha(z)|$ for all z in K_δ. (c is greater than zero because $\bar{g}_\alpha(z)$ vanishes neither inside K_δ, nor on its boundary), and let M be an upper bound for all $|\bar{g}_\nu(z)|$ in K_δ : $|\bar{g}_\nu(z)| < M$. Then, with $0 < \epsilon < \frac{1}{2}$ arbitrary to begin with, and for all z in K_δ, and all w within and on the boundary of the circle K_ϵ with center w_0 and radius ϵ, we have

$$|A| < \frac{M \cdot \epsilon}{c}(1 + \epsilon + \cdots + \epsilon^{m-\alpha-1}) < 2\frac{M}{c} \cdot \epsilon.$$

Now let ϵ be taken definitely less than $\frac{c}{4M}$, but otherwise arbitrarily, and let it remain fixed from now on. Then, for all z in K_δ, and all w in and on K_ϵ,

$$|A| = |A(z, w)| < \frac{1}{2}.$$

We now choose $0 < \delta < \delta'$ so small, that, in the interior of the circle K_δ with center z_0 and radius δ, the absolute values $|\bar{g}_0(z)|$, $|\bar{g}_1(z)|$, $\cdots$, $|\bar{g}_{\alpha-1}(z)|$ all remain less than the fixed number

$$\mu = \frac{c}{2\left(\dfrac{1}{\epsilon} + \dfrac{1}{\epsilon^2} + \cdots + \dfrac{1}{\epsilon^\alpha}\right)}.$$

[1] It is clear what is meant in the cases $\alpha = 1$ and $\alpha = m$.

(This is possible because all these coefficients vanish *at the point* z_0.) Then, for all z in K_δ and all w *on the boundary of* K_ϵ, i.e., for all z and w for which

$$|z - z_0| < \delta \quad \text{and} \quad |w - w_0| \overset{[sic!]}{=} \epsilon,$$

we have

$$|B| = |B(z, w)| < \mu \cdot \frac{1}{c}\left(\frac{1}{\epsilon} + \frac{1}{\epsilon^2} + \cdots + \frac{1}{\epsilon^\alpha}\right) = \frac{1}{2}.$$

We shall prove that our assertion is valid for these circles K_δ and K_ϵ. Let z_1 be an arbitrary point in K_δ. For all w *on the boundary* of K_ϵ,

$$|\bar{g}_\alpha(z_1) \cdot (w - w_0)^\alpha|$$

$$> |\bar{g}_\alpha(z_1) \cdot (w - w_0)^\alpha (A(z_1, w) + B(z_1, w))|,$$

because there both A and B in absolute value remain less than $\frac{1}{2}$. If we apply Rouché's theorem (cf. p. 111) to the functions of $w(!)$ inside these absolute-value signs, and to the circumference of K_ϵ, we see immediately that $G(z_1, w)$ has precisely the same number of zeros in the *interior* of the circle K_ϵ as the function $\bar{g}_\alpha(z_1) \cdot$ $\cdot (w - w_0)^\alpha$ on the left does, i.e., *precisely* α *zeros.* And these must be distinct, because at z_1, which also lies in $K_{\delta'}$, $D(z_1) \neq 0$.

Now we make the further assumption that $D(z_0) \neq 0$, i.e., that $\alpha = 1$ at z_0. Then, for every $z = z_1$ in K_δ, there is *one, and only one,* root of $G(z_1, w) = 0$ in K_ϵ. Consequently, this root is a single-valued and continuous function, $f_1(z)$, of z, concerning which we have the

Theorem on the differentiability of the roots. $w = f_1(z)$ *is a regular function of z in K_δ.*

Proof: Let z_1 be an arbitrary point, and $z_1 + \zeta$ a neighboring point, both in the interior of K_δ. Let $f_1(z_1) = w_1$ and $f_1(z_1 + \zeta) = w_1 + \omega$, so that $G(z_1, w_1) = 0$, $G(z_1 + \zeta, w_1 + \omega) = 0$, and—because of the continuity of the function $f_1(z)$—as $\zeta \to 0$, $\omega \to 0$. Our new assertion then is simply that

$$\lim_{\zeta \to 0} \frac{f_1(z_1 + \zeta) - f_1(z_1)}{\zeta} = \lim_{\zeta \to 0} \frac{\omega}{\zeta}$$

exists. Now, if we arrange in ascending powers of ζ and ω, $G(z_1 + \zeta, w_1 + \omega) = G(z_1, w_1) + \zeta \cdot G_z(z_1, w_1) + \omega \cdot G_w(z_1, w_1) + \{$terms which contain at least the factors ζ^2, $\zeta\omega$, or $\omega^2\}$. Here $G_z(z_1, w_1)$ and $G_w(z_1, w_1)$ denote, as usual, the respective (partial) derivatives of $G(z, w)$ with respect to z, w alone, at (z_1, w_1). Since the left-hand side and the first term on the right are equal to zero, we can write

$$0 = \zeta[G_z(z_1, w_1) + P \cdot \zeta + Q \cdot \omega]$$

$$+ \omega [G_w(z_1, w_1) + R \cdot \omega],$$

where, for brevity, P, Q, R denote certain entire rational functions of ζ and ω. Here $G_w(z_1, w_1) \neq 0$, since w_1 was assumed to be a simple root of $G(z_1, w) = 0$; and we can therefore suppose ζ, and with it, ω, already so small, that

$$| R \cdot \omega | < | G_w(z_1, w_1) |.$$

But then the second bracket in the last equation is not zero; and it follows immediately that

$$f_1'(z_1) = \lim_{\zeta \to 0} \frac{\omega}{\zeta} \text{ exists and is equal to } -\frac{G_z(z_1,\ w_1)}{G_w(z_1,\ w_1)},$$

Q. E. D.

§15. The Algebraic Function

The theorems of the preceding paragraph have brought about the following situation: To every non-critical point z_0 of the plane, there correspond m distinct values, which can be combined in every sufficiently small circle about such a point—briefly: "in the small" —so as to form m separate single-valued and regular functional elements, which we shall denote by

$$f_1(z; z_0),\ f_2(z; z_0),\ \cdots,\ f_m(z, z_0).$$

These may be thought of as power series with center z_0.

We have now to show that all these elements belong to one and the same m-valued analytic function.

1. We see, first of all, that each one of the elements can be continued unhindered over the punctured plane. For, let K_0 be any circle in which one of our elements, say $f_1(z, z_0)$, is regular, and let z_1 be a non-critical boundary-point of K_0. Then,—because of the uniqueness of the combination in the small,—precisely one of the elements $f_\nu(z; z_1)$, $(\nu = 1, 2, \cdots, m)$, must coincide with $f_1(z, z_0)$ in that part of the neighborhood of z_1 which lies interior to K_0, wherewith the possibility of continuing f_1 is already demonstrated.

2. We now imagine the critical points $a_1, a_2, \cdots, a_r$ to be joined in any order, and then joined to the point

∞, by a simple line, L, composed of rectilinear segments and a half-line; and the plane to be cut along L (as in Fig. 7). Then each of the ele-

ments $f_\nu(z; z_0)$ can be continued unhindered over the cut
plane—we shall denote this
simply connected region by
$\mathfrak{E}'$—so that, according to the
monodromy theorem (cf. I,
p. 105), each thus gives rise to

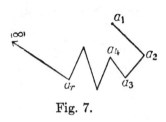

Fig. 7.

a *single-valued* and *regular* function in $\mathfrak{E}'$. We shall denote the resulting functions by $F_1(z)$, $F_2(z)$, $\cdots$, $F_m(z)$, respectively. These functions, which are obviously independent of the choice of the starting-point z_0, combine "in the large" the entire m-fold domain of values borne by the points of $\mathfrak{E}'$ to form m separate single-valued and regular functions in $\mathfrak{E}'$; and, when substituted for w, satisfy the algebraic equation $G(\mathfrak{z}, w) = 0$ for every z in $\mathfrak{E}'$. All that remains to be shown, now, is that all these m functions can be continued into one another across the boundary, L, of $\mathfrak{E}'$— in a few words, that they are the m branches of one and the same analytic function. To this end we investigate

3. the behavior of the functions at the critical points and at ∞. Let a be one of the critical points; K, the circumference of a circle about a, which neither encloses nor contains any further critical points; z_0, a point of K. Then every one of the m elements $f_\nu(z; z_0)$ can be continued along K (say in the positive sense). On returning to the point z_0, each of these elements must— again because of the uniqueness of the combination of the domain of values in the small—be continued into a definite (but, in general, *different*) one of these,—and, of course, never two distinct elements into one and the same, since otherwise the inverse continuation would transform this last element into two distinct ones. To put it briefly, the m elements thus undergo a *permutation*. We suppose the elements to be numbered in such a manner, that the permutation carries f_1 into f_2, f_2 into f_3, $\cdots$, f_{p-1} into f_p, and f_p back again into f_1, $(1 \leq p \leq m)$, so that the first p elements form a *cycle*.[1]

Then, in particular, $f_1(z; z_0)$, and with it, $F_1(z)$, goes over into itself after a p-fold continuation around a. If we accordingly set

$$(z - a) = (z')^p \quad \text{and} \quad F_1(z) = F_1(z'^p + a) = \varphi_1(z'),$$

$\varphi_1(z')$ is not only *regular*, but also *single-valued*, in a neighborhood of $z' = 0$, apart from this point itself. For, as the variable z' encircles the origin once (i.e., as its amplitude is increased continuously by 2π), z'^p encircles the origin, and hence z the point a, precisely p times, since the amplitude of $z - a$ is increased by $2p\pi$. Consequently, $\varphi_1(z')$ can be developed in a Laurent series,

$$\varphi_1(z') = \sum_{n=-\infty}^{+\infty} c_n z'^n,$$

[1]If, in a permutation of m objects, a subset of these undergoes a *"cyclic"* permutation, we say that the elements of this subset form a *cycle*. There is then the simple **Theorem:** *Every permutation can be expressed as the product of disjunct cycles.*

For example, let $m = 9$, and suppose that the figures 1, 2, 3, 4, 5, 6, 7, 8, 9 are transformed into 3, 7, 5, 4, 1, 8, 9, 6, 2, respectively. Then the figures 1, 3, 5, as well as 2, 7, 9, form a cycle of degree three; the figures 6, 8, a cycle of degree two; the figure 4 by itself, a cycle of degree one.

for a neighborhood of the origin, so that $F_1(z)$ admits of an expansion of the form

$$F_1(z) = \sum_{n=-\infty}^{+\infty} c_n \left(\sqrt[p]{z-a} \right)^n$$

for a neighborhood of the critical point a.[1] We now make the further assertion:

Only a finite number of negative powers appear in this expansion.

Proof: If $g_m(a) \neq 0$, so that $G(a, w) = 0$ has precisely m roots, some of which, however, are *multiple* roots, the theorem on the continuity of the roots states that these are continuous *at* the point a. Hence, in this case, *no* negative powers can appear in the above expansion.

But if $g_m = 0$, to the qth order, say, we must proceed otherwise. We can set $g_m(z) = (z - a)^q \cdot h_m(z)$, where $h_m(a) \neq 0$. Then, if we form

$$(z - a)^{q(m-1)} \cdot G(z, w),$$

one can verify immediately that, on setting $(z - a)^q \cdot w = v$, this can be written in the form

$$\Psi(z, v) = h_0(z) + h_1(z) \cdot v + \cdots + h_m(z) \cdot v^m,$$

where $h_0, h_1, \cdots, h_{m-1}$ denote suitable entire rational functions of z. Obviously the equation $\Psi(z, v) = 0$ is also irreducible, and, moreover, for *its* highest coefficient we have $h_m(a) \neq 0$. Hence, the roots of this new equation are continuous at $z = a$, and consequently,

---

[1] It is easy to verify that this one expansion represents all the p functions $F_1, F_2, \cdots, F_p$ of our cycle of degree p if we substitute for $\sqrt[p]{z - a}$ its p meanings. We shall not have to make use of this remark, however.

as in the case $g_m(a) \neq 0$ just treated, admit of an expansion of the form in question, in which, however, *no* negative powers appear. Since $w = (z - a)^{-q} \cdot v$, it follows immediately that the roots of the given equation, and hence our functions $F_\nu(z)$, also admit of an expansion of the same form in a neighborhood of the critical point $z = a$, and that *at most a finite number* of negative powers (namely, at most $p \cdot q$) can appear in this expansion, Q. E. D.

For the point $z = \infty$ the considerations proceed quite similarly; one has only to replace $z - a$ everywhere by $1/z$, and regard a sufficiently large circle as the circumference, K, surrounding the point ∞. These considerations, whose details everyone will be able to carry out for himself, show that each of the functions $F_\nu(z)$ admits of an expansion, for a neighborhood of the point ∞, of the form

$$\sum_{n=-\infty}^{+\infty} c_n \left(\sqrt[p]{\frac{1}{z}} \right)^n, \qquad (1 \leq p \leq m),$$

in which at most a finite number of negative powers of the pth root appear.

The critical points, with which we shall also class the point ∞, have thus been shown to be singularities of a particularly simple kind. We make the following

Definition. *If an analytic function is regular, though not necessarily single-valued, in a neighborhood of a point a or ∞,—apart from this point itself,—and if it admits of an expansion there of the form*

$$\sum_{n=-\infty}^{+\infty} c_n \left(\sqrt[p]{z - a} \right)^n, \qquad \sum_{n=-\infty}^{+\infty} c_n \left(\sqrt[p]{\frac{1}{z}} \right)^n,$$

respectively, in which only a finite number of negative powers of the pth root appear, then this point shall be called an algebraic point.[1] *We also say that the function has there the character of an algebraic function.*

4. We can now finally lay the keystone, and prove that every one of the m functions $F_\nu(z)$ can be continued into any other one by means of a suitable continuation across the cut, L.

For this purpose it is sufficient to show that F_1 can be continued into any other F_ν. For, if we can carry F_1 into F_ν and also into F_μ, then first, by the inverse continuation, F_ν is carried into F_1, and then further, in this indirect way, it passes from F_1 into F_μ; so that, in any case, F_ν can be continued into F_μ. Assume, now, that it is *impossible* to continue F_1 into some F_ν: suppose that these functions have been numbered so that F_1 can be carried into $F_2, F_3, \cdots, F_k, (k < m)$, but not into $F_{k+1}, \cdots, F_m$. This means, then, that by arbitrary continuation in the punctured plane, the first k functions are always permuted *among themselves* and are never carried into any of the remaining ones. If we form any *symmetric* function of them, $S(F_1, F_2, \cdots, F_k) = \Phi(z)$, it does not change at all, and is therefore single-valued and regular in the punctured plane. For a neighborhood of any critical point (including ∞), $\Phi(z)$ can be developed in an ordinary Laurent series having, according to **3.**, only a *finite number* of negative powers.

This means that $\Phi(z)$ has no singularities other than

[1]If $p = 1$, we are dealing with an ordinary *pole*; if, in addition to this, *no* negative powers appear, the point is actually regular. Naturally, we speak of an algebraic *singularity* only when this last is not the case.

poles in the entire plane (including ∞), and hence, according to I, §35, Theorem 2, it is a *rational* function of z. In particular,

$$(w - F_1)(w - F_2) \cdots (w - F_k)$$

$$= \varphi_0(z) + \varphi_1(z) \cdot w + \cdots + \varphi_k(z) \cdot w^k = 0$$

is an equation whose coefficients $\varphi_\lambda(z)$ are all *rational* functions of z. If we multiply this equation by a common denominator of the coefficients, there results an equation of the form

$$g(z, w) = \gamma_0(z) + \gamma_1(z) \cdot w + \cdots + \gamma_k(z) \cdot w^k = 0$$

with *entire rational* coefficients: an algebraic equation which is satisfied by the functions $F_1, F_2, \cdots, F_k$. But that is impossible for $k < m$ because of the assumed irreducibility of $G(z, w)$.[1] Our assumption is therefore untenable; and we have thus proved the theorem stated in the end of §13, and beyond that, the following

Theorem. *An algebraic function has no singularities other than algebraic singularities in the entire plane (including ∞).*

5. It is now an easy matter to construct the Riemann surface for the algebraic function defined by $G(z, w) = 0$. Corresponding to the m functions $F_\nu(z)$, we take m sheets, all cut along L, whose points bear the values of

---

[1]For, there is the purely algebraic **Theorem:** *If the equation $g(z, w) = 0$ has a root in common with the irreducible equation $G(z, w) = 0$ for all z of a region, then $G(z, w)$ is a factor of $g(z, w)$; and hence, the degree in w of g is at least as high as that of G.*

the functions F_1, F_2, $\cdots$, F_m, respectively. If we continue these functions one at a time across one of the segments of the cut, L, connecting two successive critical points, each of the F_ν goes over again into a definite one of these. We join the m sheets to one another in the manner hereby fully uniquely required,[1] whereupon this cut-segment disappears.

If we imagine the corresponding process to be carried out for *all* segments of the cut (including that which extends to ∞), *all* boundaries disappear, and the Riemann surface for the algebraic function defined by $G(z, w) = 0$ is complete. We see it more compactly, and the exceptional role of the point ∞ vanishes, if we start with the sphere instead of the plane. Then we have a closed m-sheeted Riemann sphere before us, every non-critical point of which is the bearer of one, and only one, functional value.

Finally, we shall make the critical points bearers of functional values, for which the following method suggests itself: By continuing around a critical point a (which may also be ∞), the m functions F_ν undergo, as we saw, a definite permutation, which can be decomposed into a certain number, say $l(1 \leq l \leq m)$ of disjunct cycles. Then, the point a shall be added to the surface, but counted only l (not m) times, *once* for all the sheets *together* that are connected in one and the same cycle. Every single one of these l superposed points a shall now be made bearer of the value c_0, ∞, according as *the expansion*, obtained in **3.**, *which corre-*

[1]Some sheets, in particular, may pass "smoothly" over the cut-segment,—if, namely, the function in question, F_ν, is carried into itself in crossing the cut.

sponds to it begins with the constant term c_0 or actually contains negative powers.[1]

Now that we have enlarged the domain of values in this manner, we call the totality of pairs of values (z, w), consisting of all points z of our Riemann sphere as first component, and the functional values w uniquely corresponding to these points as second component, the **algebraic configuration** *defined by* $G(z, w) = 0$. Its further, exhaustive investigation forms the subject of the theory of algebraic functions.

Exercise. Discuss in detail the structure of the Riemann surfaces (critical points; method of joining the sheets, and behavior of the function, at those points; distribution of the domain of values; etc.) for the algebraic functions, w, of z, defined by

a) $w^3 - 1 - z = 0,$

b) $w^3 - 3w - z = 0,$

c) $w + \dfrac{1}{w} - z = 0.$

---

[1]If the l cycles, in turn, are of degree $p_1, p_2, \cdots , p_l$, then exactly l branch-points, of order $p_1 - 1, p_2 - 1, \cdots , p_l - 1,$ respectively, are superposed at a, (among which, in particular, branch-points of order zero, i.e., ordinary points, may also appear); and these, counted as l distinct points of the surface, may of course bear entirely different functional values.

CHAPTER 6

THE ANALYTIC CONFIGURATION

§16. The Monogenic Analytic Function

We are now in a position to supplement the definition of the complete analytic function which was given in I, pp. 102-103 but which still contains several omissions, and thereby give a certain completeness to our investigations, at least with respect to the fundamental idea—that of the analytic function. To this end we resume the considerations of §10.

We started there with a given functional element—say a power series—and continued it as long as possible. We must now indicate somewhat more precisely how this is to be carried out, be it only purely theoretically. For we shall require, in general, an infinite number of power series before a further continuation leads to nothing new. If, however, one is to give a constructive procedure according to which the continuation can be carried out in its entirety, it must consist of only an *enumerable number* of steps. This appears impossible at first, because it would seem that to exhaust the continuation possibilities of even only the first power series, one would have to form a new expansion about *every* point of its circle of convergence as center. But then there would be a *non-enumerable infinity* of new power series.

It is easy to see, however, that in a continuation

process—let us say the continuation of a power series
$\mathfrak{P}_0$, with center z_0, along the path k to ζ—we need only
use such new power series as have centers with *rational
coordinates*.[1] For, if the continuation along k is at all
possible, the requisite circles of convergence with centers $z_0, z_1, \cdots, z_{m-1}, z_m = \zeta$ (cf. I, p. 88 and Fig. 5)
cover a region whose boundary is a positive distance,
ρ, from k. If, now, we employ, instead of $z_1, z_2, \cdots$, any
rational centers $z'_1, z'_2, \cdots$, each having a distance of at
most $\tfrac{1}{2}\rho$ from k, we also arrive at ζ, and with the same
functional element.

The rational points form an enumerable set, and from
this we are able to infer that the entire continuation
process for a functional element can be completed in an
enumerable number of steps. For, only an enumerable
number of new power series, say

$$\mathfrak{P}_{01}, \mathfrak{P}_{02}, \cdots, \mathfrak{P}_{0n}, \cdots,$$

result from the given power series $\mathfrak{P}_0$ if we make merely
the rational points of its circle of convergence centers
of the new expansions. At most an enumerable infinity
arise again from each of these, so that we obtain, on the
whole, only an enumerable number of *new* power series,[2]
say

$$\mathfrak{P}_{11}, \mathfrak{P}_{12}, \cdots, \mathfrak{P}_{1n}, \cdots.$$

For these the argument repeats itself, etc.; so that we
get all in all an enumerable number of sequences of

[1] For brevity we shall call such points *rational points*.

[2] That an enumerable set of enumerable sets of objects is itself
an enumerable set of these objects was proved when we arranged
the lattice points of the plane in a sequence; see p. 28.

enumerably many power series, and hence *all together
an enumerable number of such series*, which we shall
denote finally by

$$\mathfrak{P}_0, \ \mathfrak{P}_1, \ \mathfrak{P}_2, \ \cdots, \ \mathfrak{P}_r, \ \cdots .$$

This proves

Theorem 1. *If it is at all possible to include an arbitrary
point z, by means of (power-series) continuation of the
initial element $\mathfrak{P}_0$ along some path, in the interior of the
circle of convergence of a new power series, it can be
effected with the exclusive use of (in each case a finite
number of) power series of a suitably fixed sequence
$\mathfrak{P}_1, \mathfrak{P}_2, \cdots, \mathfrak{P}_r, \cdots$ of such.*

Suppose that we have exhausted in this manner all
possibilities of continuing a given functional element
$w = f(z)$. The result of this is that a neighborhood of
every point z_0 of the plane which appears at all in the
interior of one of the circles of convergence of the $\mathfrak{P}_r$
receives a finite or an enumerably infinite number of
different coverings with functional values in such a
manner, that every single covering forms a single-valued
and regular function in a neighborhood of z_0. Let these
be the functions

$$f_1(z; z_0), \ f_2(z; z_0), \ \cdots .$$

We then let the point z_0 bear the values of these func-
tions for $z = z_0$; denote these values by $w_0^{(1)}$, $w_0^{(2)}$, $\cdots$.
If the same value should appear more than once in this
process, it shall be borne by z_0 correspondingly often.
Finally, if we imagine the pairs of numbers

$$(z_0, \ w_0^{(1)}), \ (z_0, \ w_0^{(2)}), \ \cdots$$

to be formed for *every* z_0 belonging to the interior of at least one of the circles of convergence of the $\mathfrak{P}_r$, these pairs in their totality constitute the **monogenic analytic function** *generated by the initial element.* The function is thus determined by the following properties:

1. To every point z of the plane, or of a part of it, there correspond a finite or an enumerably infinite number of functional values $w^{(1)}$, $w^{(2)}$, $\cdots$ (among which the same ones may appear in an arbitrary manner).

2. If (z_0, w_0) is a particular one of these pairs of values, the *totality* of pairs (z, w), whose first component belongs to a neighborhood of z_0, can be combined so as to form a finite or an enumerably infinite number of regular functions $f_r(z, z_0)$ at z_0.

3. If $f_r(z; z_0)$ and $f_\mu(z; z_0)$ are an arbitrary pair of the functions thus formed, each is an (of course not *immediate*) continuation of the other.

4. If any one of these functional elements $f_r(z; z_0)$ is developed in a power series with a rational center, we obtain one of the power series $\mathfrak{P}_r$.

Accordingly, we can state, in particular, the following two theorems:

Theorem 2. *Every domain of values which is given in any manner "in the small"[1] generates, if at all, precisely one well-determined monogenic analytic function.*

Theorem 3. *The set of functional values which a*

[1] I.e., every covering of a region, however small, (or of a path segment, or of only a bounded infinite set of points) of the z-plane with w-values (cf. in this connection the considerations in I, p. 95).

multiple-valued function can assume at a point z_0 is either finite or enumerably infinite.

§17. The Riemann Surface

There is nothing now to prevent the construction of the Ricmann surface belonging to a monogenic analytic function, according to the procedure indicated in §10: corresponding to the sequence of the $\mathfrak{P}_r$, we paste the disks of their circles of convergence together in the manner there described—penetrating (in imagination) intermediate sheets if necessary—and thus obtain the required surface.[1]

Naturally, the way in which the sheets are joined together *may* become very complicated. It *may* also, however, be very clear and transparent, as the examples treated in chapters 4 and 5 show. The Riemann surface lays no claim to being an end in itself, but is only intended as an aid to the imagination. One will therefore leave it aside in all those cases in which the joining of the sheets becomes so involved, that it would be more difficult to follow the functional values on the surface than with the function itself. Thus, the advantage of constructing, e.g., the surface for $w = \text{arc sin } z$ in order to visualize the course of this function is no longer worth mentioning, although it would be very simple to set it up[2] on the basis of, say, the formula $w = \text{arc sin } z$

---

[1] Instead of circular disks, we may, of course, take any other regions; in particular, such *maximal* regions in which a branch of the function remains regular. Thus, e.g., for the algebraic functions we could take the entire cut plane at once.

[2] It is nevertheless quite useful to construct these surfaces in imagination in order to get practice in using the ideas involved.

$= -i \log (iz + \sqrt{1 - z^2})$, which is obtained from $z = \sin w = -\frac{1}{2}i(e^{iw} - e^{-iw})$ by solving for w. But, e.g., in the case of the inverse of Weierstrass's σ- or $\wp$-function, the construction of the corresponding surface will offer hardly any advantage any more.

It is therefore not advisable to continue the formation of the Riemann surface in the most general instance. One should rather see from case to case whether its construction helps perception or not. We have become acquainted with the most important examples of useful surfaces in the preceding two chapters. As far as the general case is concerned, it is sufficient to know that, for a given function, a Riemann surface can be constructed at all events, on which its values form a *single-valued* function of position. Every point z is covered by as many (a finite or an infinite number of) sheets (see below) as there are different elements for a neighborhood of this point, and these sheets hang together in a perfectly definite manner. This last means that if we begin at a certain point z_0 of a particular sheet and describe any definite path (more precisely: a path whose projection on the ordinary z-plane is given), its course on the surface is fully unique, and consequently, leads us to a perfectly definite point of a perfectly definite sheet,—provided only that the path does not leave the surface, i.e., provided that it avoids the singular boundary points of those sheets (see below) on which it lies.

We are now finally in a position to formulate more precisely several concepts which we have already made much use of:

1. A **sheet** of the Riemann surface is obtained if,

starting with any one of our circular disks, we paste on new disks (or parts thereof), according to the above-described procedure, so long, but only so long, as we do not get a multiple covering of the plane. The concept of the sheet is thus, as we particularly emphasize, not an absolute one, but depends on the execution of the construction procedure just described. Nevertheless, it has a well-determined sense to speak of the different sheets on which a particular point z_0 lies: z_0 lies on as many different sheets as the number of times it is an interior point of distinct circular disks (i.e., disks not pasted together at z_0 and a neighborhood thereof). The totality of points z which belong to one and the same sheet form a region in the sense of I, §4.

2. By a **branch** of a given (multiple-valued) analytic function $F(z)$ we mean any function which is represented by the covering of *one sheet* of the proper Riemann surface, and which is single-valued and analytic in the region corresponding to it by 1.

3. By a **functional element** of an analytic function $F(z)$ we mean the representation of any branch or of only a part of it; in particular, each of the power series $\mathfrak{P}_r$, and each of the functions $f_r(z; z_0)$ used in §16,—for which, moreover, one can imagine the boundary of the neighborhood of z_0 which comes into question to be fixed in various ways,—is a functional element.

4. The concept of the **singular point** is, like that of the branch or the sheet, not an absolute one either: a particular point can be called singular or regular only for a certain branch or a certain sheet (cf. the example on p. 96). For this, however, the concept is fully determined. For, the region which, according to 1., is

filled by the totality of points z belonging to a sheet is covered with a domain of values which, by 2., forms there a single-valued analytic function—the branch belonging to this sheet. For this function the boundary points of the region in question are divided (cf. I, §24) unambiguously into regular and singular points, i.e., those at which the continuation across the boundary is possible, and those at which it is impossible, respectively.

Exercises. 1. Discuss in detail the structure of the Riemann surfaces for the functions

a) $w = z^a$ (a complex, arbitrary),

b) $w = $ arc sin z.

(Cf. §15, ex. 2c.)

2. Construct a function for which the unit circle is the natural boundary, but which is

a) exactly two-valued, b) infinitely multiple-valued in the interior of the unit circle.

§18. The Analytic Configuration

We have yet to take up a last small supplement (similar to that which we made in the conclusion of §15 in the case of the algebraic functions), by means of which, then, the notion of a complete analytic function becomes settled in every respect.

The state of affairs thus far is the following: the domain of values which finds itself affixed to a neighborhood of an (*eo ipso*: *interior*) point of a sheet of the Riemann surface forms there a regular functional element, whereas all singular points of the separate branches (sheets) are, at first, not added to the surface at all. Among these singular points there are some of such a simple nature, that it is—also for various other reasons—advantageous to class them, so to speak, with the regular points, or, in any case, to add them to the surface. These are, in a few words, the algebraic singularities,—namely, the following points:

1. *The poles on a sheet*; i.e., every isolated boundary point z_0 of a sheet, such that the domain of values attached to a neighborhood of z_0 can be developed in an (ordinary) Laurent series with *only a finite number* of negative powers.[1] We let such a point bear the value ∞, and we add the pair (z_0, ∞) to the number pairs of the monogenic analytic function.

2. *The algebraic branch-points*; i.e., every singular boundary point of one of the sheets, about which a finite number, say $p(>1)$, distinct sheets hang together like the surface for $\sqrt[p]{z}$ at the origin, and for which the following condition is fulfilled: the domain of values affixed to these p sheets in a neighborhood of z_0, which (cf. p. 129) at all events can be developed in a series of the form

$$\sum_{n=-\infty}^{+\infty} c_n \left(\sqrt[p]{z - z_0} \right)^n ,$$

shall be such, that no negative powers of $\sqrt[p]{z - z_0}$, or only a finite number of these, appear in this expansion.

We shall add such a point to the surface, and count it *once* for these p sheets *together*. We let it bear the value ∞ or c_0, according as negative powers do or do not appear in the expansion, and we add the pair

---

[1] On some other sheet, z_0 may very well be a regular point, or a different kind of singular point.

(z_0, ∞), (z_0, c_0), respectively, once to our pairs of numbers (z, w).

3. Finally, we shall add *the point* ∞ to the surface under corresponding conditions, namely, in a few words, if the behavior at the point ∞, *when regarded on the sphere*, is the same as that at the point z_0 in the cases just considered; in detail, if either

a) a certain sheet is simple in a neighborhood of the point ∞ [1], and the domain of values attached to it there forms a single-valued regular function whose Laurent expansion about ∞ contains at most a finite number of negative powers of $\left(\dfrac{1}{z}\right)$;

or

b) a finite number, say $p(>1)$, of distinct sheets hang together about the point ∞ like the surface for $\sqrt[p]{z}$ about this point, and the (at all events possible) development of the affixed domain of values in the series

$$\sum_{n=-\infty}^{+\infty} c_n \left(\sqrt[p]{\frac{1}{z}} \right)^n$$

contains at most a *finite number* of negative powers of [2] $\sqrt[p]{\dfrac{1}{z}}$.

In case a) we say that there is an ordinary point, in case b), that there is a *branch-point of order* $p - 1$, at

[1] I.e., the sheet in question contains *all* points z which lie in the exterior of a certain circle.

[2] The cases 1 and 3a can, of course, be interpreted as the special cases of 2 and 3b obtained when $p = 1$.

the point ∞. It shall be added to the surface in both cases, and counted *precisely once* for the p sheets together that were taken into consideration. We let it bear the value ∞ or c_0, according as negative powers do or do not appear in the expansion in question. We accordingly add the pair (∞, ∞), (∞, c_0), respectively, once to our pairs of numbers.

We say, now, that the set of pairs (z, w), which has been supplemented in this way, represents **the (monogenic) analytic configuration** defined by the initial element.[1]

It is useful to add to the set of our functional elements $f_r(z; z_0)$ the finite or enumerably infinite number of expansions which we spoke of in 1. $-$ 3. Then we have before us in the set of all these functional elements or in the set of all our pairs of numbers (z, w), completely and in clear arrangement, the *configuration* which arises, in the continuation process, from an arbitrarily given power series or other representation of a regular function in the small.

In conclusion let us add that the *theory of uniformization* mentioned on p. 84, footnote, is, in a way, the connecting link between the two main subjects of our

[1]Without proof we add the remark that, by interchanging the two components of every number pair of an analytic configuration (z, w) which arises from a functional element $w = f(z)$, there results another monogenic analytic configuration (w, z) which is designated as the *inverse configuration*. This transparent transition from an analytic function to its inverse could not be formulated so simply and clearly without the supplements met with in this paragraph. Their usefulness is already sufficiently assured by this fact alone.

investigation, the single-valued and the multiple-valued functions. For in it is proved the theorem that *any (multiple-valued) analytic function $w = F(z)$ can be completely represented (uniformized) with the aid of single-valued functions*; and this more precisely in the sense that there always exist two single-valued functions of the complex variable t, $z = z(t)$ and $w = w(t)$, with the property that the pair $(z, w) = (z(t), w(t))$ yields the complete analytic function $w = F(z)$ when the variable t runs over a certain domain of its plane. (General uniformization theorem of Poincaré and Koebe.)

BIBLIOGRAPHY

In addition to the works dealing with the general theory which were mentioned in *Theory of Functions, Part I*, we call attention to the following books relating to the more special classes of functions considered in the present volume:

P. Appell and É. Goursat, *Théorie des Fonctions Algébriques et de leurs Intégrales*, 2 vols., 2d edition, Paris, 1929-1930.

É. Borel, *Leçons sur les Fonctions Entières*, 2d edition, Paris, 1921.
————, *Leçons sur les Fonctions Méromorphes*, Paris, 1903.

H. Burkhardt, *Funktionentheoretische Vorlesungen*, vol. II, 3d edition, edited by G. Faber, Berlin, 1920.

H. Durège, *Theorie der elliptischen Funktionen*, 5th edition, revised by L. Maurer, Leipzig, 1908.

R. Fricke, *Die elliptischen Funktionen und ihre Anwendungen*, part I, Leipzig and Berlin, 1916, part II, Leipzig, 1922.

H. Hancock, *Lectures on the Theory of Elliptic Functions*, vol. I, New York, 1910.

K. Hensel and G. Landsberg, *Theorie der algebraischen Funktionen einer Variabeln und ihre Anwendung auf algebraische Kurven und Abelsche Integrale*, Leipzig, 1902.

C. Jordan, *Cours d'Analyse*, vol. II, 3d edition, Paris, 1913.

E. T. Whittaker and G. N. Watson, *A Course of Modern Analysis,* New York, 1945.

INDEX

A CATALOG OF
SELECTED DOVER BOOKS
IN ALL FIELDS OF INTEREST

A CATALOG OF SELECTED DOVER
BOOKS IN ALL FIELDS OF INTEREST

CONCERNING THE SPIRITUAL IN ART, Wassily Kandinsky. Pioneering work by father of abstract art. Thoughts on color theory, nature of art. Analysis of earlier masters. 12 illustrations. 80pp. of text. 5⅜ × 8½. 23411-8 Pa. $2.50

LEONARDO ON THE HUMAN BODY, Leonardo da Vinci. More than 1200 of Leonardo's anatomical drawings on 215 plates. Leonardo's text, which accompanies the drawings, has been translated into English. 506pp. 8⅜ × 11¾.
24483-0 Pa. $10.95

GOBLIN MARKET, Christina Rossetti. Best-known work by poet comparable to Emily Dickinson, Alfred Tennyson. With 46 delightfully grotesque illustrations by Laurence Housman. 64pp. 4 × 6¾. 24516-0 Pa. $2.50

THE HEART OF THOREAU'S JOURNALS, edited by Odell Shepard. Selections from *Journal*, ranging over full gamut of interests. 228pp. 5⅜ × 8½.
20741-2 Pa. $4.50

MR. LINCOLN'S CAMERA MAN: MATHEW B. BRADY, Roy Meredith. Over 300 Brady photos reproduced directly from original negatives, photos. Lively commentary. 368pp. 8⅜ × 11¼. 23021-X Pa. $11.95

PHOTOGRAPHIC VIEWS OF SHERMAN'S CAMPAIGN, George N. Barnard. Reprint of landmark 1866 volume with 61 plates: battlefield of New Hope Church, the Etawah Bridge, the capture of Atlanta, etc. 80pp. 9 × 12. 23445-2 Pa. $6.00

A SHORT HISTORY OF ANATOMY AND PHYSIOLOGY FROM THE GREEKS TO HARVEY, Dr. Charles Singer. Thoroughly engrossing non-technical survey. 270 illustrations. 211pp. 5⅜ × 8½. 20389-1 Pa. $4.50

REDOUTE ROSES IRON-ON TRANSFER PATTERNS, Barbara Christopher. Redouté was botanical painter to the Empress Josephine; transfer his famous roses onto fabric with these 24 transfer patterns. 80pp. 8¼ × 10⅞. 24292-7 Pa. $3.50

THE FIVE BOOKS OF ARCHITECTURE, Sebastiano Serlio. Architectural milestone, first (1611) English translation of Renaissance classic. Unabridged reproduction of original edition includes over 300 woodcut illustrations. 416pp. 9⅜ × 12¼. 24349-4 Pa. $14.95

CARLSON'S GUIDE TO LANDSCAPE PAINTING, John F. Carlson. Authoritative, comprehensive guide covers, every aspect of landscape painting. 34 reproductions of paintings by author; 58 explanatory diagrams. 144pp. 8⅜ × 11.
22927-0 Pa. $4.95

101 PUZZLES IN THOUGHT AND LOGIC, C.R. Wylie, Jr. Solve murders, robberies, see which fishermen are liars—purely by reasoning! 107pp. 5⅜ × 8½.
20367-0 Pa. $2.00

TEST YOUR LOGIC, George J. Summers. 50 more truly new puzzles with new turns of thought, new subtleties of inference. 100pp. 5⅜ × 8½. 22877-0 Pa. $2.25

THE MURDER BOOK OF J.G. REEDER, Edgar Wallace. Eight suspenseful stories by bestselling mystery writer of 20s and 30s. Features the donnish Mr. J.G. Reeder of Public Prosecutor's Office. 128pp. 5⅜ × 8½. (Available in U.S. only)
24374-5 Pa. $3.50

ANNE ORR'S CHARTED DESIGNS, Anne Orr. Best designs by premier needlework designer, all on charts: flowers, borders, birds, children, alphabets, etc. Over 100 charts, 10 in color. Total of 40pp. 8¼ × 11.
23704-4 Pa. $2.25

BASIC CONSTRUCTION TECHNIQUES FOR HOUSES AND SMALL BUILDINGS SIMPLY EXPLAINED, U.S. Bureau of Naval Personnel. Grading, masonry, woodworking, floor and wall framing, roof framing, plastering, tile setting, much more. Over 675 illustrations. 568pp. 6½ × 9¼.
20242-9 Pa. $8.95

MATISSE LINE DRAWINGS AND PRINTS, Henri Matisse. Representative collection of female nudes, faces, still lifes, experimental works, etc., from 1898 to 1948. 50 illustrations. 48pp. 8⅜ × 11¼.
23877-6 Pa. $2.50

HOW TO PLAY THE CHESS OPENINGS, Eugene Znosko-Borovsky. Clear, profound examinations of just what each opening is intended to do and how opponent can counter. Many sample games. 147pp. 5⅜ × 8½.
22795-2 Pa. $2.95

DUPLICATE BRIDGE, Alfred Sheinwold. Clear, thorough, easily followed account: rules, etiquette, scoring, strategy, bidding; Goren's point-count system, Blackwood and Gerber conventions, etc. 158pp. 5⅜ × 8½.
22741-3 Pa. $3.00

SARGENT PORTRAIT DRAWINGS, J.S. Sargent. Collection of 42 portraits reveals technical skill and intuitive eye of noted American portrait painter, John Singer Sargent. 48pp. 8¼ × 11⅛.
24524-1 Pa. $2.95

ENTERTAINING SCIENCE EXPERIMENTS WITH EVERYDAY OBJECTS, Martin Gardner. Over 100 experiments for youngsters. Will amuse, astonish, teach, and entertain. Over 100 illustrations. 127pp. 5⅜ × 8½.
24201-3 Pa. $2.50

TEDDY BEAR PAPER DOLLS IN FULL COLOR: A Family of Four Bears and Their Costumes, Crystal Collins. A family of four Teddy Bear paper dolls and nearly 60 cut-out costumes. Full color, printed one side only. 32pp. 9¼ × 12¼.
24550-0 Pa. $3.50

NEW CALLIGRAPHIC ORNAMENTS AND FLOURISHES, Arthur Baker. Unusual, multi-useable material: arrows, pointing hands, brackets and frames, ovals, swirls, birds, etc. Nearly 700 illustrations. 80pp. 8⅜ × 11¼.
24095-9 Pa. $3.75

DINOSAUR DIORAMAS TO CUT & ASSEMBLE, M. Kalmenoff. Two complete three-dimensional scenes in full color, with 31 cut-out animals and plants. Excellent educational toy for youngsters. Instructions; 2 assembly diagrams. 32pp. 9¼ × 12¼.
24541-1 Pa. $3.95

SILHOUETTES: A PICTORIAL ARCHIVE OF VARIED ILLUSTRATIONS, edited by Carol Belanger Grafton. Over 600 silhouettes from the 18th to 20th centuries. Profiles and full figures of men, women, children, birds, animals, groups and scenes, nature, ships, an alphabet. 144pp. 8⅜ × 11¼.
23781-8 Pa. $4.95

25 KITES THAT FLY, Leslie Hunt. Full, easy-to-follow instructions for kites made from inexpensive materials. Many novelties. 70 illustrations. 110pp. 5⅜ × 8½.
22550-X Pa. $2.25

PIANO TUNING, J. Cree Fischer. Clearest, best book for beginner, amateur. Simple repairs, raising dropped notes, tuning by easy method of flattened fifths. No previous skills needed. 4 illustrations. 201pp. 5⅜ × 8½. 23267-0 Pa. $3.50

EARLY AMERICAN IRON-ON TRANSFER PATTERNS, edited by Rita Weiss. 75 designs, borders, alphabets, from traditional American sources. 48pp. 8¼ × 11.
23162-3 Pa. $1.95

CROCHETING EDGINGS, edited by Rita Weiss. Over 100 of the best designs for these lovely trims for a host of household items. Complete instructions, illustrations. 48pp. 8¼ × 11. 24031-2 Pa. $2.25

FINGER PLAYS FOR NURSERY AND KINDERGARTEN, Emilie Poulsson. 18 finger plays with music (voice and piano); entertaining, instructive. Counting, nature lore, etc. Victorian classic. 53 illustrations. 80pp. 6½ × 9¼. 22588-7 Pa. $1.95

BOSTON THEN AND NOW, Peter Vanderwarker. Here in 59 side-by-side views are photographic documentations of the city's past and present. 119 photographs. Full captions. 122pp. 8¼ × 11. 24312-5 Pa. $6.95

CROCHETING BEDSPREADS, edited by Rita Weiss. 22 patterns, originally published in three instruction books 1939-41. 39 photos, 8 charts. Instructions. 48pp. 8¼ × 11. 23610-2 Pa. $2.00

HAWTHORNE ON PAINTING, Charles W. Hawthorne. Collected from notes taken by students at famous Cape Cod School; hundreds of direct, personal *apercus*, ideas, suggestions. 91pp. 5⅜ × 8½. 20653-X Pa. $2.50

THERMODYNAMICS, Enrico Fermi. A classic of modern science. Clear, organized treatment of systems, first and second laws, entropy, thermodynamic potentials, etc. Calculus required. 160pp. 5⅜ × 8½. 60361-X Pa. $4.00

TEN BOOKS ON ARCHITECTURE, Vitruvius. The most important book ever written on architecture. Early Roman aesthetics, technology, classical orders, site selection, all other aspects. Morgan translation. 331pp. 5⅜ × 8½. 20645-9 Pa. $5.50

THE CORNELL BREAD BOOK, Clive M. McCay and Jeanette B. McCay. Famed high-protein recipe incorporated into breads, rolls, buns, coffee cakes, pizza, pie crusts, more. Nearly 50 illustrations. 48pp. 8¼ × 11. 23995-0 Pa. $2.00

THE CRAFTSMAN'S HANDBOOK, Cennino Cennini. 15th-century handbook, school of Giotto, explains applying gold, silver leaf; gesso; fresco painting, grinding pigments, etc. 142pp. 6⅛ × 9¼. 20054-X Pa. $3.50

FRANK LLOYD WRIGHT'S FALLINGWATER, Donald Hoffmann. Full story of Wright's masterwork at Bear Run, Pa. 100 photographs of site, construction, and details of completed structure. 112pp. 9¼ × 10. 23671-4 Pa. $6.50

OVAL STAINED GLASS PATTERN BOOK, C. Eaton. 60 new designs framed in shape of an oval. Greater complexity, challenge with sinuous cats, birds, mandalas framed in antique shape. 64pp. 8¼ × 11. 24519-5 Pa. $3.50

THE BOOK OF WOOD CARVING, Charles Marshall Sayers. Still finest book for beginning student. Fundamentals, technique; gives 34 designs, over 34 projects for panels, bookends, mirrors, etc. 33 photos. 118pp. 7¾ × 10⅝. 23654-4 Pa. $3.95

CARVING COUNTRY CHARACTERS, Bill Higginbotham. Expert advice for beginning, advanced carvers on materials, techniques for creating 18 projects—mirthful panorama of American characters. 105 illustrations. 80pp. 8⅜ × 11. 24135-1 Pa. $2.50

300 ART NOUVEAU DESIGNS AND MOTIFS IN FULL COLOR, C.B. Grafton. 44 full-page plates display swirling lines and muted colors typical of Art Nouveau. Borders, frames, panels, cartouches, dingbats, etc. 48pp. 9⅜ × 12¼. 24354-0 Pa. $6.00

SELF-WORKING CARD TRICKS, Karl Fulves. Editor of *Pallbearer* offers 72 tricks that work automatically through nature of card deck. No sleight of hand needed. Often spectacular. 42 illustrations. 113pp. 5⅜ × 8½. 23334-0 Pa. $3.50

CUT AND ASSEMBLE A WESTERN FRONTIER TOWN, Edmund V. Gillon, Jr. Ten authentic full-color buildings on heavy cardboard stock in H-O scale. Sheriff's Office and Jail, Saloon, Wells Fargo, Opera House, others. 48pp. 9¼ × 12¼. 23736-2 Pa. $3.95

CUT AND ASSEMBLE AN EARLY NEW ENGLAND VILLAGE, Edmund V. Gillon, Jr. Printed in full color on heavy cardboard stock. 12 authentic buildings in H-O scale: Adams home in Quincy, Mass., Oliver Wight house in Sturbridge, smithy, store, church, others. 48pp. 9¼ × 12¼. 23536-X Pa. $3.95

THE TALE OF TWO BAD MICE, Beatrix Potter. Tom Thumb and Hunca Munca squeeze out of their hole and go exploring. 27 full-color Potter illustrations. 59pp. 4¼ × 5½. (Available in U.S. only) 23065-1 Pa. $1.50

CARVING FIGURE CARICATURES IN THE OZARK STYLE, Harold L. Enlow. Instructions and illustrations for ten delightful projects, plus general carving instructions. 22 drawings and 47 photographs altogether. 39pp. 8⅜ × 11. 23151-8 Pa. $2.50

A TREASURY OF FLOWER DESIGNS FOR ARTISTS, EMBROIDERERS AND CRAFTSMEN, Susan Gaber. 100 garden favorites lushly rendered by artist for artists, craftsmen, needleworkers. Many form frames, borders. 80pp. 8¼ × 11. 24096-7 Pa. $3.50

CUT & ASSEMBLE A TOY THEATER/THE NUTCRACKER BALLET, Tom Tierney. Model of a complete, full-color production of Tchaikovsky's classic. 6 backdrops, dozens of characters, familiar dance sequences. 32pp. 9⅜ × 12¼. 24194-7 Pa. $4.50

ANIMALS: 1,419 COPYRIGHT-FREE ILLUSTRATIONS OF MAMMALS, BIRDS, FISH, INSECTS, ETC., edited by Jim Harter. Clear wood engravings present, in extremely lifelike poses, over 1,000 species of animals. 284pp. 9 × 12. 23766-4 Pa. $9.95

MORE HAND SHADOWS, Henry Bursill. For those at their 'finger ends,'' 16 more effects—Shakespeare, a hare, a squirrel, Mr. Punch, and twelve more—each explained by a full-page illustration. Considerable period charm. 30pp. 6½ × 9¼. 21384-6 Pa. $1.95

CATALOG OF DOVER BOOKS

SURREAL STICKERS AND UNREAL STAMPS, William Rowe. 224 haunting, hilarious stamps on gummed, perforated stock, with images of elephants, geisha girls, George Washington, etc. 16pp. one side. 8¼ × 11. 24371-0 Pa. $3.50

GOURMET KITCHEN LABELS, Ed Sibbett, Jr. 112 full-color labels (4 copies each of 28 designs). Fruit, bread, other culinary motifs. Gummed and perforated. 16pp. 8¼ × 11. 24087-8 Pa. $2.95

PATTERNS AND INSTRUCTIONS FOR CARVING AUTHENTIC BIRDS, H.D. Green. Detailed instructions, 27 diagrams, 85 photographs for carving 15 species of birds so life-like, they'll seem ready to fly! 8¼ × 11. 24222-6 Pa. $2.75

FLATLAND, E.A. Abbott. Science-fiction classic explores life of 2-D being in 3-D world. 16 illustrations. 103pp. 5⅜ × 8. 20001-9 Pa. $2.00

DRIED FLOWERS, Sarah Whitlock and Martha Rankin. Concise, clear, practical guide to dehydration, glycerinizing, pressing plant material, and more. Covers use of silica gel. 12 drawings. 32pp. 5⅜ × 8½. 21802-3 Pa. $1.00

EASY-TO-MAKE CANDLES, Gary V. Guy. Learn how easy it is to make all kinds of decorative candles. Step-by-step instructions. 82 illustrations. 48pp. 8¼ × 11. 23881-4 Pa. $2.50

SUPER STICKERS FOR KIDS, Carolyn Bracken. 128 gummed and perforated full-color stickers: GIRL WANTED, KEEP OUT, BORED OF EDUCATION, X-RATED, COMBAT ZONE, many others. 16pp. 8¼ × 11. 24092-4 Pa. $2.50

CUT AND COLOR PAPER MASKS, Michael Grater. Clowns, animals, funny faces...simply color them in, cut them out, and put them together, and you have 9 paper masks to play with and enjoy. 32pp. 8¼ × 11. 23171-2 Pa. $2.25

A CHRISTMAS CAROL: THE ORIGINAL MANUSCRIPT, Charles Dickens. Clear facsimile of Dickens manuscript, on facing pages with final printed text. 8 illustrations by John Leech, 4 in color on covers. 144pp. 8⅜ × 11¼. 20980-6 Pa. $5.95

CARVING SHOREBIRDS, Harry V. Shourds & Anthony Hillman. 16 full-size patterns (all double-page spreads) for 19 North American shorebirds with step-by-step instructions. 72pp. 9¼ × 12¼. 24287-0 Pa. $4.95

THE GENTLE ART OF MATHEMATICS, Dan Pedoe. Mathematical games, probability, the question of infinity, topology, how the laws of algebra work, problems of irrational numbers, and more. 42 figures. 143pp. 5⅜ × 8½. (EBE) 22949-1 Pa. $3.50

READY-TO-USE DOLLHOUSE WALLPAPER, Katzenbach & Warren, Inc. Stripe, 2 floral stripes, 2 allover florals, polka dot; all in full color. 4 sheets (350 sq. in.) of each, enough for average room. 48pp. 8¼ × 11. 23495-9 Pa. $2.95

MINIATURE IRON-ON TRANSFER PATTERNS FOR DOLLHOUSES, DOLLS, AND SMALL PROJECTS, Rita Weiss and Frank Fontana. Over 100 miniature patterns: rugs, bedspreads, quilts, chair seats, etc. In standard dollhouse size. 48pp. 8¼ × 11. 23741-9 Pa. $1.95

THE DINOSAUR COLORING BOOK, Anthony Rao. 45 renderings of dinosaurs, fossil birds, turtles, other creatures of Mesozoic Era. Scientifically accurate. Captions. 48pp. 8¼ × 11. 24022-3 Pa. $2.25

JAPANESE DESIGN MOTIFS, Matsuya Co. Mon, or heraldic designs. Over 4000 typical, beautiful designs: birds, animals, flowers, swords, fans, geometrics; all beautifully stylized. 213pp. 11⅜ × 8¼. 22874-6 Pa. $7.95

THE TALE OF BENJAMIN BUNNY, Beatrix Potter. Peter Rabbit's cousin coaxes him back into Mr. McGregor's garden for a whole new set of adventures. All 27 full-color illustrations. 59pp. 4¼ × 5½. (Available in U.S. only) 21102-9 Pa. $1.50

THE TALE OF PETER RABBIT AND OTHER FAVORITE STORIES BOXED SET, Beatrix Potter. Seven of Beatrix Potter's best-loved tales including Peter Rabbit in a specially designed, durable boxed set. 4¼ × 5½. Total of 447pp. 158 color illustrations. (Available in U.S. only) 23903-9 Pa. $10.80

PRACTICAL MENTAL MAGIC, Theodore Annemann. Nearly 200 astonishing feats of mental magic revealed in step-by-step detail. Complete advice on staging, patter, etc. Illustrated. 320pp. 5⅜ × 8½. 24426-1 Pa. $5.95

CELEBRATED CASES OF JUDGE DEE (DEE GOONG AN), translated by Robert Van Gulik. Authentic 18th-century Chinese detective novel; Dee and associates solve three interlocked cases. Led to van Gulik's own stories with same characters. Extensive introduction. 9 illustrations. 237pp. 5⅜ × 8½. 23337-5 Pa. $4.50

CUT & FOLD EXTRATERRESTRIAL INVADERS THAT FLY, M. Grater. Stage your own lilliputian space battles.By following the step-by-step instructions and explanatory diagrams you can launch 22 full-color fliers into space. 36pp. 8¼ × 11. 24478-4 Pa. $2.95

CUT & ASSEMBLE VICTORIAN HOUSES, Edmund V. Gillon, Jr. Printed in full color on heavy cardboard stock, 4 authentic Victorian houses in H-O scale: Italian-style Villa, Octagon, Second Empire, Stick Style. 48pp. 9¼ × 12¼. 23849-0 Pa. $3.95

BEST SCIENCE FICTION STORIES OF H.G. WELLS, H.G. Wells. Full novel *The Invisible Man,* plus 17 short stories: "The Crystal Egg," "Aepyornis Island," "The Strange Orchid," etc. 303pp. 5⅜ × 8½. (Available in U.S. only) 21531-8 Pa. $4.95

TRADEMARK DESIGNS OF THE WORLD, Yusaku Kamekura. A lavish collection of nearly 700 trademarks, the work of Wright, Loewy, Klee, Binder, hundreds of others. 160pp. 8¾ × 8. (Available in U.S. only) 24191-2 Pa. $5.00

THE ARTIST'S AND CRAFTSMAN'S GUIDE TO REDUCING, ENLARGING AND TRANSFERRING DESIGNS, Rita Weiss. Discover, reduce, enlarge, transfer designs from any objects to any craft project. 12pp. plus 16 sheets special graph paper. 8¼ × 11. 24142-4 Pa. $3.25

TREASURY OF JAPANESE DESIGNS AND MOTIFS FOR ARTISTS AND CRAFTSMEN, edited by Carol Belanger Grafton. Indispensable collection of 360 traditional Japanese designs and motifs redrawn in clean, crisp black-and-white, copyright-free illustrations. 96pp. 8¼ × 11. 24435-0 Pa. $3.95

CATALOG OF DOVER BOOKS

CHANCERY CURSIVE STROKE BY STROKE, Arthur Baker. Instructions and illustrations for each stroke of each letter (upper and lower case) and numerals. 54 full-page plates. 64pp. 8¼ × 11. 24278-1 Pa. $2.50

THE ENJOYMENT AND USE OF COLOR, Walter Sargent. Color relationships, values, intensities; complementary colors, illumination, similar topics. Color in nature and art. 7 color plates, 29 illustrations. 274pp. 5⅜ × 8½. 20944-X Pa. $4.50

SCULPTURE PRINCIPLES AND PRACTICE, Louis Slobodkin. Step-by-step approach to clay, plaster, metals, stone; classical and modern. 253 drawings, photos. 255pp. 8⅛ × 11. 22960-2 Pa. $7.50

VICTORIAN FASHION PAPER DOLLS FROM HARPER'S BAZAR, 1867-1898, Theodore Menten. Four female dolls with 28 elegant high fashion costumes, printed in full color. 32pp. 9¼ × 12¼. 23453-3 Pa. $3.50

FLOPSY, MOPSY AND COTTONTAIL: A Little Book of Paper Dolls in Full Color, Susan LaBelle. Three dolls and 21 costumes (7 for each doll) show Peter Rabbit's siblings dressed for holidays, gardening, hiking, etc. Charming borders, captions. 48pp. 4¼ × 5½. 24376-1 Pa. $2.25

NATIONAL LEAGUE BASEBALL CARD CLASSICS, Bert Randolph Sugar. 83 big-leaguers from 1909-69 on facsimile cards. Hubbell, Dean, Spahn, Brock plus advertising, info, no duplications. Perforated, detachable. 16pp. 8¼ × 11. 24308-7 Pa. $2.95

THE LOGICAL APPROACH TO CHESS, Dr. Max Euwe, et al. First-rate text of comprehensive strategy, tactics, theory for the amateur. No gambits to memorize, just a clear, logical approach. 224pp. 5⅜ × 8½. 24353-2 Pa. $4.50

MAGICK IN THEORY AND PRACTICE, Aleister Crowley. The summation of the thought and practice of the century's most famous necromancer, long hard to find. Crowley's best book. 436pp. 5⅜ × 8½. (Available in U.S. only) 23295-6 Pa. $6.50

THE HAUNTED HOTEL, Wilkie Collins. Collins' last great tale; doom and destiny in a Venetian palace. Praised by T.S. Eliot. 127pp. 5⅜ × 8½. 24333-8 Pa. $3.00

ART DECO DISPLAY ALPHABETS, Dan X. Solo. Wide variety of bold yet elegant lettering in handsome Art Deco styles. 100 complete fonts, with numerals, punctuation, more. 104pp. 8⅛ × 11. 24372-9 Pa. $4.00

CALLIGRAPHIC ALPHABETS, Arthur Baker. Nearly 150 complete alphabets by outstanding contemporary. Stimulating ideas; useful source for unique effects. 154 plates. 157pp. 8⅜ × 11¼. 21045-6 Pa. $4.95

ARTHUR BAKER'S HISTORIC CALLIGRAPHIC ALPHABETS, Arthur Baker. From monumental capitals of first-century Rome to humanistic cursive of 16th century, 33 alphabets in fresh interpretations. 88 plates. 96pp. 9 × 12. 24054-1 Pa. $4.50

LETTIE LANE PAPER DOLLS, Sheila Young. Genteel turn-of-the-century family very popular then and now. 24 paper dolls. 16 plates in full color. 32pp. 9¼ × 12¼. 24089-4 Pa. $3.50

CATALOG OF DOVER BOOKS

KEYBOARD WORKS FOR SOLO INSTRUMENTS, G.F. Handel. 35 neglected works from Handel's vast oeuvre, originally jotted down as improvisations. Includes Eight Great Suites, others. New sequence. 174pp. 9⅜ × 12¼.
24338-9 Pa. $7.50

AMERICAN LEAGUE BASEBALL CARD CLASSICS, Bert Randolph Sugar. 82 stars from 1900s to 60s on facsimile cards. Ruth, Cobb, Mantle, Williams, plus advertising, info, no duplications. Perforated, detachable. 16pp. 8¼ × 11.
24286-2 Pa. $2.95

A TREASURY OF CHARTED DESIGNS FOR NEEDLEWORKERS, Georgia Gorham and Jeanne Warth. 141 charted designs: owl, cat with yarn, tulips, piano, spinning wheel, covered bridge, Victorian house and many others. 48pp. 8¼ × 11.
23558-0 Pa. $1.95

DANISH FLORAL CHARTED DESIGNS, Gerda Bengtsson. Exquisite collection of over 40 different florals: anemone, Iceland poppy, wild fruit, pansies, many others. 45 illustrations. 48pp. 8¼ × 11.
23957-8 Pa. $1.75

OLD PHILADELPHIA IN EARLY PHOTOGRAPHS 1839-1914, Robert F. Looney. 215 photographs: panoramas, street scenes, landmarks, President-elect Lincoln's visit, 1876 Centennial Exposition, much more. 230pp. 8⅜ × 11¾.
23345-6 Pa. $9.95

PRELUDE TO MATHEMATICS, W.W. Sawyer. Noted mathematician's lively, stimulating account of non-Euclidean geometry, matrices, determinants, group theory, other topics. Emphasis on novel, striking aspects. 224pp. 5⅜ × 8½.
24401-6 Pa. $4.50

ADVENTURES WITH A MICROSCOPE, Richard Headstrom. 59 adventures with clothing fibers, protozoa, ferns and lichens, roots and leaves, much more. 142 illustrations. 232pp 5⅜ × 8½.
23471-1 Pa. $3.95

IDENTIFYING ANIMAL TRACKS: MAMMALS, BIRDS, AND OTHER ANIMALS OF THE EASTERN UNITED STATES, Richard Headstrom. For hunters, naturalists, scouts, nature lovers. Diagrams of tracks, tips on identification. 128pp. 5⅜ × 8.
24442-3 Pa. $3.50

VICTORIAN FASHIONS AND COSTUMES FROM HARPER'S BAZAR, 1867-1898, edited by Stella Blum. Day costumes, evening wear, sports clothes, shoes, hats, other accessories in over 1,000 detailed engravings. 320pp. 9⅜ × 12¼.
22990-4 Pa. $9.95

EVERYDAY FASHIONS OF THE TWENTIES AS PICTURED IN SEARS AND OTHER CATALOGS, edited by Stella Blum. Actual dress of the Roaring Twenties, with text by Stella Blum. Over 750 illustrations, captions. 156pp. 9 × 12.
24134-3 Pa. $8.50

HALL OF FAME BASEBALL CARDS, edited by Bert Randolph Sugar. Cy Young, Ted Williams, Lou Gehrig, and many other Hall of Fame greats on 92 full-color, detachable reprints of early baseball cards. No duplication of cards with *Classic Baseball Cards.* 16pp. 8¼ × 11.
23624-2 Pa. $3.50

THE ART OF HAND LETTERING, Helm Wotzkow. Course in hand lettering, Roman, Gothic, Italic, Block, Script. Tools, proportions, optical aspects, individual variation. Very quality conscious. Hundreds of specimens. 320pp. 5⅜ × 8½.
21797-3 Pa. $4.95

HOW THE OTHER HALF LIVES, Jacob A. Riis. Journalistic record of filth, degradation, upward drive in New York immigrant slums, shops, around 1900. New edition includes 100 original Riis photos, monuments of early photography. 233pp. 10 × 7⅞. 22012-5 Pa. $7.95

CHINA AND ITS PEOPLE IN EARLY PHOTOGRAPHS, John Thomson. In 200 black-and-white photographs of exceptional quality photographic pioneer Thomson captures the mountains, dwellings, monuments and people of 19th-century China. 272pp. 9⅜ × 12¼. 24393-1 Pa. $12.95

GODEY COSTUME PLATES IN COLOR FOR DECOUPAGE AND FRAMING, edited by Eleanor Hasbrouk Rawlings. 24 full-color engravings depicting 19th-century Parisian haute couture. Printed on one side only. 56pp. 8¼ × 11. 23879-2 Pa. $3.95

ART NOUVEAU STAINED GLASS PATTERN BOOK, Ed Sibbett, Jr. 104 projects using well-known themes of Art Nouveau: swirling forms, florals, peacocks, and sensuous women. 60pp. 8¼ × 11. 23577-7 Pa. $3.50

QUICK AND EASY PATCHWORK ON THE SEWING MACHINE: Susan Aylsworth Murwin and Suzzy Payne. Instructions, diagrams show exactly how to machine sew 12 quilts. 48pp. of templates. 50 figures. 80pp. 8¼ × 11. 23770-2 Pa. $3.50

THE STANDARD BOOK OF QUILT MAKING AND COLLECTING, Marguerite Ickis. Full information, full-sized patterns for making 46 traditional quilts, also 150 other patterns. 483 illustrations. 273pp. 6⅞ × 9⅝. 20582-7 Pa. $5.95

LETTERING AND ALPHABETS, J. Albert Cavanagh. 85 complete alphabets lettered in various styles; instructions for spacing, roughs, brushwork. 121pp. 8¾ × 8. 20053-1 Pa. $3.75

LETTER FORMS: 110 COMPLETE ALPHABETS, Frederick Lambert. 110 sets of capital letters; 16 lower case alphabets; 70 sets of numbers and other symbols. 110pp. 8⅛ × 11. 22872-X Pa. $4.50

ORCHIDS AS HOUSE PLANTS, Rebecca Tyson Northen. Grow cattleyas and many other kinds of orchids—in a window, in a case, or under artificial light. 63 illustrations. 148pp. 5⅜ × 8½. 23261-1 Pa. $2.95

THE MUSHROOM HANDBOOK, Louis C.C. Krieger. Still the best popular handbook. Full descriptions of 259 species, extremely thorough text, poisons, folklore, etc. 32 color plates; 126 other illustrations. 560pp. 5⅜ × 8½. 21861-9 Pa. $8.50

THE DORÉ BIBLE ILLUSTRATIONS, Gustave Doré. All wonderful, detailed plates: Adam and Eve, Flood, Babylon, life of Jesus, etc. Brief King James text with each plate. 241 plates. 241pp. 9 × 12. 23004-X Pa. $8.95

THE BOOK OF KELLS: Selected Plates in Full Color, edited by Blanche Cirker. 32 full-page plates from greatest manuscript-icon of early Middle Ages. Fantastic, mysterious. Publisher's Note. Captions. 32pp. 9⅜ × 12¼. 24345-1 Pa. $4.50

THE PERFECT WAGNERITE, George Bernard Shaw. Brilliant criticism of the Ring Cycle, with provocative interpretation of politics, economic theories behind the Ring. 136pp. 5⅜ × 8½. (Available in U.S. only) 21707-8 Pa. $3.00

THE RIME OF THE ANCIENT MARINER, Gustave Doré, S.T. Coleridge. Doré's finest work, 34 plates capture moods, subtleties of poem. Full text. 77pp. 9¼ × 12. 22305-1 Pa. $4.95

SONGS OF INNOCENCE, William Blake. The first and most popular of Blake's famous "Illuminated Books," in a facsimile edition reproducing all 31 brightly colored plates. Additional printed text of each poem. 64pp. 5¼ × 7. 22764-2 Pa. $3.00

AN INTRODUCTION TO INFORMATION THEORY, J.R. Pierce. Second (1980) edition of most impressive non-technical account available. Encoding, entropy, noisy channel, related areas, etc. 320pp. 5⅜ × 8½. 24061-4 Pa. $4.95

THE DIVINE PROPORTION: A STUDY IN MATHEMATICAL BEAUTY, H.E. Huntley. "Divine proportion" or "golden ratio" in poetry, Pascal's triangle, philosophy, psychology, music, mathematical figures, etc. Excellent bridge between science and art. 58 figures. 185pp. 5⅜ × 8½. 22254-3 Pa. $3.95

THE DOVER NEW YORK WALKING GUIDE: From the Battery to Wall Street, Mary J. Shapiro. Superb inexpensive guide to historic buildings and locales in lower Manhattan: Trinity Church, Bowling Green, more. Complete Text; maps. 36 illustrations. 48pp. 3⅞ × 9¼. 24225-0 Pa. $2.50

NEW YORK THEN AND NOW, Edward B. Watson, Edmund V. Gillon, Jr. 83 important Manhattan sites: on facing pages early photographs (1875-1925) and 1976 photos by Gillon. 172 illustrations. 171pp. 9¼ × 10. 23361-8 Pa. $7.95

HISTORIC COSTUME IN PICTURES, Braun & Schneider. Over 1450 costumed figures from dawn of civilization to end of 19th century. English captions. 125 plates. 256pp. 8⅜ × 11¼. 23150-X Pa. $7.50

VICTORIAN AND EDWARDIAN FASHION: A Photographic Survey, Alison Gernsheim. First fashion history completely illustrated by contemporary photographs. Full text plus 235 photos, 1840-1914, in which many celebrities appear. 240pp. 6½ × 9¼. 24205-6 Pa. $6.00

CHARTED CHRISTMAS DESIGNS FOR COUNTED CROSS-STITCH AND OTHER NEEDLECRAFTS, Lindberg Press. Charted designs for 45 beautiful needlecraft projects with many yuletide and wintertime motifs. 48pp. 8¼ × 11. 24356-7 Pa. $1.95

101 FOLK DESIGNS FOR COUNTED CROSS-STITCH AND OTHER NEEDLE-CRAFTS, Carter Houck. 101 authentic charted folk designs in a wide array of lovely representations with many suggestions for effective use. 48pp. 8¼ × 11. 24369-9 Pa. $2.25

FIVE ACRES AND INDEPENDENCE, Maurice G. Kains. Great back-to-the-land classic explains basics of self-sufficient farming. The one book to get. 95 illustrations. 397pp. 5⅜ × 8½. 20974-1 Pa. $4.95

A MODERN HERBAL, Margaret Grieve. Much the fullest, most exact, most useful compilation of herbal material. Gigantic alphabetical encyclopedia, from aconite to zedoary, gives botanical information, medical properties, folklore, economic uses, and much else. Indispensable to serious reader. 161 illustrations. 888pp. 6½ × 9¼. (Available in U.S. only) 22798-7, 22799-5 Pa., Two-vol. set $16.45

DECORATIVE NAPKIN FOLDING FOR BEGINNERS, Lillian Oppenheimer and Natalie Epstein. 22 different napkin folds in the shape of a heart, clown's hat, love knot, etc. 63 drawings. 48pp. 8¼ × 11. 23797-4 Pa. $1.95

DECORATIVE LABELS FOR HOME CANNING, PRESERVING, AND OTHER HOUSEHOLD AND GIFT USES, Theodore Menten. 128 gummed, perforated labels, beautifully printed in 2 colors. 12 versions. Adhere to metal, glass, wood, ceramics. 24pp. 8¼ × 11. 23219-0 Pa. $2.95

EARLY AMERICAN STENCILS ON WALLS AND FURNITURE, Janet Waring. Thorough coverage of 19th-century folk art: techniques, artifacts, surviving specimens. 166 illustrations, 7 in color. 147pp. of text. 7⅞ × 10¾. 21906-2 Pa. $9.95

AMERICAN ANTIQUE WEATHERVANES, A.B. & W.T. Westervelt. Extensively illustrated 1883 catalog exhibiting over 550 copper weathervanes and finials. Excellent primary source by one of the principal manufacturers. 104pp. 6⅛ × 9¼. 24396-6 Pa. $3.95

ART STUDENTS' ANATOMY, Edmond J. Farris. Long favorite in art schools. Basic elements, common positions, actions. Full text, 158 illustrations. 159pp. 5⅜ × 8½. 20744-7 Pa. $3.95

BRIDGMAN'S LIFE DRAWING, George B. Bridgman. More than 500 drawings and text teach you to abstract the body into its major masses. Also specific areas of anatomy. 192pp. 6½ × 9¼. (EA) 22710-3 Pa. $4.50

COMPLETE PRELUDES AND ETUDES FOR SOLO PIANO, Frederic Chopin. All 26 Preludes, all 27 Etudes by greatest composer of piano music. Authoritative Paderewski edition. 224pp. 9 × 12. (Available in U.S. only) 24052-5 Pa. $7.50

PIANO MUSIC 1888-1905, Claude Debussy. Deux Arabesques, Suite Bergamesque, Masques, 1st series of Images, etc. 9 others, in corrected editions. 175pp. 9⅜ × 12¼. (ECE) 22771-5 Pa. $5.95

TEDDY BEAR IRON-ON TRANSFER PATTERNS, Ted Menten. 80 iron-on transfer patterns of male and female Teddys in a wide variety of activities, poses, sizes. 48pp. 8¼ × 11. 24596-9 Pa. $2.25

A PICTURE HISTORY OF THE BROOKLYN BRIDGE, M.J. Shapiro. Profusely illustrated account of greatest engineering achievement of 19th century. 167 rare photos & engravings recall construction, human drama. Extensive, detailed text. 122pp. 8¼ × 11. 24403-2 Pa. $7.95

NEW YORK IN THE THIRTIES, Berenice Abbott. Noted photographer's fascinating study shows new buildings that have become famous and old sights that have disappeared forever. 97 photographs. 97pp. 11⅜ × 10. 22967-X Pa. $6.50

MATHEMATICAL TABLES AND FORMULAS, Robert D. Carmichael and Edwin R. Smith. Logarithms, sines, tangents, trig functions, powers, roots, reciprocals, exponential and hyperbolic functions, formulas and theorems. 269pp. 5⅜ × 8½. 60111-0 Pa. $3.75

HANDBOOK OF MATHEMATICAL FUNCTIONS WITH FORMULAS, GRAPHS, AND MATHEMATICAL TABLES, edited by Milton Abramowitz and Irene A. Stegun. Vast compendium: 29 sets of tables, some to as high as 20 places. 1,046pp. 8 × 10½. 61272-4 Pa. $19.95

REASON IN ART, George Santayana. Renowned philosopher's provocative, seminal treatment of basis of art in instinct and experience. Volume Four of *The Life of Reason*. 230pp. 5⅜ × 8.   24358-3 Pa. $4.50

LANGUAGE, TRUTH AND LOGIC, Alfred J. Ayer. Famous, clear introduction to Vienna, Cambridge schools of Logical Positivism. Role of philosophy, elimination of metaphysics, nature of analysis, etc. 160pp. 5⅜ × 8½. (USCO)
20010-8 Pa. $2.75

BASIC ELECTRONICS, U.S. Bureau of Naval Personnel. Electron tubes, circuits, antennas, AM, FM, and CW transmission and receiving, etc. 560 illustrations. 567pp. 6½ × 9¼.   21076-6 Pa. $8.95

THE ART DECO STYLE, edited by Theodore Menten. Furniture, jewelry, metalwork, ceramics, fabrics, lighting fixtures, interior decors, exteriors, graphics from pure French sources. Over 400 photographs. 183pp. 8⅜ × 11¼.
22824-X Pa. $6.95

THE FOUR BOOKS OF ARCHITECTURE, Andrea Palladio. 16th-century classic covers classical architectural remains, Renaissance revivals, classical orders, etc. 1738 Ware English edition. 216 plates. 110pp. of text. 9½ × 12¾.
21308-0 Pa. $11.50

THE WIT AND HUMOR OF OSCAR WILDE, edited by Alvin Redman. More than 1000 ripostes, paradoxes, wisecracks: Work is the curse of the drinking classes, I can resist everything except temptations, etc. 258pp. 5⅜ × 8½. (USCO)
20602-5 Pa. $3.50

THE DEVIL'S DICTIONARY, Ambrose Bierce. Barbed, bitter, brilliant witticisms in the form of a dictionary. Best, most ferocious satire America has produced. 145pp. 5⅜ × 8½.   20487-1 Pa. $2.50

ERTÉ'S FASHION DESIGNS, Erté. 210 black-and-white inventions from *Harper's Bazar*, 1918-32, plus 8pp. full color covers. Captions. 88pp. 9 × 12.
24203-X Pa. $6.50

ERTÉ GRAPHICS, Erté. Collection of striking color graphics: *Seasons, Alphabet, Numerals, Aces* and *Precious Stones*. 50 plates, including 4 on covers. 48pp. 9⅜ × 12¼.   23580-7 Pa. $6.95

PAPER FOLDING FOR BEGINNERS, William D. Murray and Francis J. Rigney. Clearest book for making origami sail boats, roosters, frogs that move legs, etc. 40 projects. More than 275 illustrations. 94pp. 5⅜ × 8½.   20713-7 Pa. $2.25

ORIGAMI FOR THE ENTHUSIAST, John Montroll. Fish, ostrich, peacock, squirrel, rhinoceros, Pegasus, 19 other intricate subjects. Instructions. Diagrams. 128pp. 9 × 12.   23799-0 Pa. $4.95

CROCHETING NOVELTY POT HOLDERS, edited by Linda Macho. 64 useful, whimsical pot holders feature kitchen themes, animals, flowers, other novelties. Surprisingly easy to crochet. Complete instructions. 48pp. 8¼ × 11.
24296-X Pa. $1.95

CROCHETING DOILIES, edited by Rita Weiss. Irish Crochet, Jewel, Star Wheel, Vanity Fair and more. Also luncheon and console sets, runners and centerpieces. 51 illustrations. 48pp. 8¼ × 11.   23424-X Pa. $2.00

YUCATAN BEFORE AND AFTER THE CONQUEST, Diego de Landa. Only significant account of Yucatan written in the early post-Conquest era. Translated by William Gates. Over 120 illustrations. 162pp. 5⅜ × 8½.   23622-6 Pa. $3.50

ORNATE PICTORIAL CALLIGRAPHY, E.A. Lupfer. Complete instructions, over 150 examples help you create magnificent "flourishes" from which beautiful animals and objects gracefully emerge. 8⅛ × 11.   21957-7 Pa. $2.95

DOLLY DINGLE PAPER DOLLS, Grace Drayton. Cute chubby children by same artist who did Campbell Kids. Rare plates from 1910s. 30 paper dolls and over 100 outfits reproduced in full color. 32pp. 9¼ × 12¼.   23711-7 Pa. $3.50

CURIOUS GEORGE PAPER DOLLS IN FULL COLOR, H. A. Rey, Kathy Allert. Naughty little monkey-hero of children's books in two doll figures, plus 48 full-color costumes: pirate, Indian chief, fireman, more. 32pp. 9¼ × 12¼.   24386-9 Pa. $3.50

GERMAN: HOW TO SPEAK AND WRITE IT, Joseph Rosenberg. Like *French, How to Speak and Write It.* Very rich modern course, with a wealth of pictorial material. 330 illustrations. 384pp. 5⅜ × 8½. (USUKO)   20271-2 Pa. $4.75

CATS AND KITTENS: 24 Ready-to-Mail Color Photo Postcards, D. Holby. Handsome collection; feline in a variety of adorable poses. Identifications. 12pp. on postcard stock. 8¼ × 11.   24469-5 Pa. $2.95

MARILYN MONROE PAPER DOLLS, Tom Tierney. 31 full-color designs on heavy stock, from *The Asphalt Jungle, Gentlemen Prefer Blondes*, 22 others. 1 doll. 16 plates. 32pp. 9⅜ × 12¼.   23769-9 Pa. $3.50

FUNDAMENTALS OF LAYOUT, F.H. Wills. All phases of layout design discussed and illustrated in 121 illustrations. Indispensable as student's text or handbook for professional. 124pp. 8⅛.× 11.   21279-3 Pa. $4.50

FANTASTIC SUPER STICKERS, Ed Sibbett, Jr. 75 colorful pressure-sensitive stickers. Peel off and place for a touch of pizzazz: clowns, penguins, teddy bears, etc. Full color. 16pp. 8¼ × 11.   24471-7 Pa. $2.95

LABELS FOR ALL OCCASIONS, Ed Sibbett, Jr. 6 labels each of 16 different designs—baroque, art nouveau, art deco, Pennsylvania Dutch, etc.—in full color. 24pp. 8¼ × 11.   23688-9 Pa. $2.95

HOW TO CALCULATE QUICKLY: RAPID METHODS IN BASIC MATHE-MATICS, Henry Sticker. Addition, subtraction, multiplication, division, checks, etc. More than 8000 problems, solutions. 185pp. 5 × 7¼.   20295-X Pa. $2.95

THE CAT COLORING BOOK, Karen Baldauski. Handsome, realistic renderings of 40 splendid felines, from American shorthair to exotic types. 44 plates. Captions. 48pp. 8¼ × 11.   24011-8 Pa. $2.25

THE TALE OF PETER RABBIT, Beatrix Potter. The inimitable Peter's terrifying adventure in Mr. McGregor's garden, with all 27 wonderful, full-color Potter illustrations. 55pp. 4¼ × 5½. (Available in U.S. only)   22827-4 Pa. $1.60

BASIC ELECTRICITY, U.S. Bureau of Naval Personnel. Batteries, circuits, conductors, AC and DC, inductance and capacitance, generators, motors, trans-formers, amplifiers, etc. 349 illustrations. 448pp. 6½ × 9¼.   20973-3 Pa. $7.95